Tucholsky Wagner Zola Scott Sydow Freud Schlegel
Turgenev Wallace Fonatne
Twain Walther von der Vogelweide Fouqué Friedrich II. von Preußen
Weber Freiligrath Frey
Fechner Fichte Weiße Rose von Fallersleben Kant Ernst Richthofen Frommel
Hölderlin
Engels Fielding Eichendorff Tacitus Dumas
Fehrs Faber Flaubert Eliasberg Zweig Ebner Eschenbach
Feuerbach Maximilian I. von Habsburg Fock Eliot Vergil
Goethe Ewald London
Elisabeth von Österreich
Mendelssohn Balzac Shakespeare Dostojewski Ganghofer
Trackl Stevenson Lichtenberg Rathenau Doyle Gjellerup
Mommsen Thoma Tolstoi Lenz Hambruch Droste-Hülshoff
Dach Verne von Arnim Hägele Hanrieder
Karrillon Reuter Rousseau Hauff Humboldt
Garschin Hagen Hauptmann Gautier
Damaschke Defoe Hebbel Baudelaire
Descartes Hegel Kussmaul Herder
Wolfram von Eschenbach Darwin Dickens Schopenhauer Rilke George
Bronner Melville Grimm Jerome Bebel Proust
Campe Horváth Aristoteles Voltaire Federer
Bismarck Vigny Barlach Heine Herodot
Gengenbach
Storm Casanova Tersteegen Grillparzer Georgy
Chamberlain Lessing Langbein Gilm Gryphius
Brentano Lafontaine
Strachwitz Claudius Schiller Kralik Iffland Sokrates
Katharina II. von Rußland Bellamy Schilling
Gerstäcker Raabe Gibbon Tschechow
Löns Hesse Hoffmann Gogol Wilde Gleim Vulpius
Luther Heym Hofmannsthal Klee Hölty Morgenstern Goedicke
Roth Heyse Klopstock Puschkin Homer Kleist
Luxemburg La Roche Horaz Mörike Musil
Machiavelli Kierkegaard Kraft Kraus
Navarra Aurel Musset Lamprecht Kind Kirchhoff Hugo Moltke
Nestroy Marie de France
Nietzsche Nansen Laotse Ipsen Liebknecht
Marx Lassalle Gorki Klett Leibniz Ringelnatz
von Ossietzky May vom Stein Lawrence Irving
Petalozzi Platon Pückler Michelangelo Knigge Kafka
Sachs Poe Liebermann Kock
Korolenko
de Sade Praetorius Mistral Zetkin

The publishing house tredition has created the series **TREDITION CLASSICS**. It contains classical literature works from over two thousand years. Most of these titles have been out of print and off the bookstore shelves for decades.

The book series is intended to preserve the cultural legacy and to promote the timeless works of classical literature. As a reader of a **TREDITION CLASSICS** book, the reader supports the mission to save many of the amazing works of world literature from oblivion.

The symbol of **TREDITION CLASSICS** is Johannes Gutenberg (1400 – 1468), the inventor of movable type printing.

With the series, tredition intends to make thousands of international literature classics available in printed format again – worldwide.

All books are available at book retailers worldwide in paperback and in hardcover. For more information please visit: www.tredition.com

tredition was established in 2006 by Sandra Latusseck and Soenke Schulz. Based in Hamburg, Germany, tredition offers publishing solutions to authors and publishing houses, combined with worldwide distribution of printed and digital book content. tredition is uniquely positioned to enable authors and publishing houses to create books on their own terms and without conventional manufacturing risks.

For more information please visit: www.tredition.com

Flying Machines: construction and operation; a practical book which shows, in illustrations, working plans and text, how to build and navigate the modern airship

William James Jackman

Imprint

This book is part of the TREDITION CLASSICS series.

Author: William James Jackman
Cover design: toepferschumann, Berlin (Germany)

Publisher: tredition GmbH, Hamburg (Germany)
ISBN: 978-3-8491-7030-1

www.tredition.com
www.tredition.de

Copyright:
The content of this book is sourced from the public domain.

The intention of the TREDITION CLASSICS series is to make world literature in the public domain available in printed format. Literary enthusiasts and organizations worldwide have scanned and digitally edited the original texts. tredition has subsequently formatted and redesigned the content into a modern reading layout. Therefore, we cannot guarantee the exact reproduction of the original format of a particular historic edition. Please also note that no modifications have been made to the spelling, therefore it may differ from the orthography used today.

PREFACE.

This book is written for the guidance of the novice in aviation—the man who seeks practical information as to the theory, construction and operation of the modern flying machine. With this object in view the wording is intentionally plain and non-technical. It contains some propositions which, so far as satisfying the experts is concerned, might doubtless be better stated in technical terms, but this would defeat the main purpose of its preparation. Consequently, while fully aware of its shortcomings in this respect, the authors have no apologies to make.

In the stating of a technical proposition so it may be clearly understood by people not versed in technical matters it becomes absolutely necessary to use language much different from that which an expert would employ, and this has been done in this volume.

No man of ordinary intelligence can read this book without obtaining a clear, comprehensive knowledge of flying machine construction and operation. He will learn, not only how to build, equip, and manipulate an aeroplane in actual flight, but will also gain a thorough understanding of the principle upon which the suspension in the air of an object much heavier than the air is made possible.

This latter feature should make the book of interest even to those who have no intention of constructing or operating a flying machine. It will enable them to better understand and appreciate the performances of the daring men like the Wright brothers, Curtiss, Bleriot, Farman, Paulhan, Latham, and others, whose bold experiments have made aviation an actuality.

For those who wish to engage in the fascinating pastime of construction and operation it is intended as a reliable, practical guide.

It may be well to explain that the sub-headings in the articles by Mr. Chanute were inserted by the authors without his knowledge. The purpose of this was merely to preserve uniformity in the typography of the book. This explanation is made in justice to Mr. Chanute.

THE AUTHORS.

IN MEMORIAM.

Octave Chanute, "the father of the modern flying machine," died at his home in Chicago on November 23, 1910, at the age of 72 years. His last work in the interest of aviation was to furnish the introductory chapter to the first edition of this volume, and to render valuable assistance in the handling of the various subjects. He even made the trip from his home to the office of the publishers one inclement day last spring, to look over the proofs of the book and, at his suggestion, several important changes were made. All this was "a labor of love" on Mr. Chanute's part. He gave of his time and talents freely because he was enthusiastic in the cause of aviation, and because he knew the authors of this book and desired to give them material aid in the preparation of the work—a favor that was most sincerely appreciated.

The authors desire to make acknowledgment of many courtesies in the way of valuable advice, information, etc., extended by Mr. Octave Chanute, C. E., Mr. E. L. Jones, Editor of Aeronautics, and the publishers of, the New England Automobile Journal and Fly.

CONTENTS

PREFACE.
IN MEMORIAM.

CHAPTER I. EVOLUTION OF TWO-SURFACE FLYING MACHINE.
CHAPTER II. THEORY, DEVELOPMENT, AND USE.
CHAPTER III. MECHANICAL BIRD ACTION
CHAPTER IV. VARIOUS FORMS OF FLYING MACHINES.
CHAPTER V. CONSTRUCTING A GLIDING MACHINE.
CHAPTER VI. LEARNING TO FLY.
CHAPTER VII. PUTTING ON THE RUDDER.
CHAPTER VIII. THE REAL FLYING MACHINE.
CHAPTER IX. SELECTION OF THE MOTOR.
CHAPTER X. PROPER DIMENSIONS OF MACHINES.
CHAPTER XI. PLANE AND RUDDER CONTROL.
CHAPTER XII. HOW TO USE THE MACHINE.
CHAPTER XIII. PECULIARITIES OF AIRSHIP POWER.
CHAPTER XIV. ABOUT WIND CURRENTS, ETC.
CHAPTER XV. THE ELEMENT OF DANGER.
CHAPTER XVI. RADICAL CHANGES BEING MADE.
CHAPTER XVII. SOME OF THE NEW DESIGNS.
CHAPTER XVIII. DEMAND FOR FLYING MACHINES.
CHAPTER XIX. LAW OF THE AIRSHIP.
CHAPTER XX. SOARING FLIGHT.
CHAPTER XXI. FLYING MACHINES VS. BALLOONS.
CHAPTER XXII. PROBLEMS OF AERIAL FLIGHT.

CHAPTER XXIII. AMATEURS MAY USE WRIGHT PATENTS.
CHAPTER XXIV. HINTS ON PROPELLER CONSTRUCTION.
CHAPTER XXV. NEW MOTORS AND DEVICES.
CHAPTER XXVI. MONOPLANES, TRIPLANES, MULTIPLANES.
CHAPTER XXVII. 1911 AEROPLANE RECORDS.
NOTABLE CROSS-COUNTRY FLIGHTS OF 1911.
CHAPTER XXVIII. GLOSSARY OF AERONAUTICAL TERMS.

Footnotes:

FLYING MACHINES: CONSTRUCTION and OPERATION

CHAPTER I. EVOLUTION OF TWO-SURFACE FLYING MACHINE.

By Octave Chanute.

I am asked to set forth the development of the "two-surface" type of flying machine which is now used with modifications by Wright Brothers, Farman, 1 Delagrange, Herring and others.

This type originated with Mr. F. H. Wenham, who patented it in England in 1866 (No. 1571), taking out provisional papers only. In the abridgment of British patent Aeronautical Specifications (1893) it is described as follows:

"Two or more aeroplanes are arranged one above the other, and support a framework or car containing the motive power. The aeroplanes are made of silk or canvas stretched on a frame by wooden rods or steel ribs. When manual power is employed the body is placed horizontally, and oars or propellers are actuated by the arms or legs.

"A start may be obtained by lowering the legs and running down hill or the machine may be started from a moving carriage. One or more screw propellers may be applied for propelling when steam power is employed."

On June 27, 1866, Mr. Wenham read before the "Aeronautical Society of Great Britain," then recently organized, the ablest paper ever presented to that society, and thereby breathed into it a spirit which has continued to this day. In this paper he described his observations of birds, discussed the laws governing flight as to the surfaces and power required both with wings and screws, and he then gave an account of his own experiments with models and with aeroplanes of sufficient size to carry the weight of a man.

Second Wenham Aeroplane.

His second aeroplane was sixteen feet from tip to tip. A trussed spar at the bottom carried six superposed bands of thin holland fabric fifteen inches wide, connected with vertical webs of holland two feet apart, thus virtually giving a length of wing of ninety-six feet and one hundred and twenty square feet of supporting surface. The man was placed horizontally on a base board beneath the spar. This apparatus when tried in the wind was found to be unmanageable by reason of the fluttering motions of the fabric, which was insufficiently stiffened with crinoline steel, but Mr. Wenham pointed out that this in no way invalidated the principle of the apparatus, which was to obtain large supporting surfaces without increasing unduly the leverage and consequent weight of spar required, by simply superposing the surfaces.

This principle is entirely sound and it is surprising that it is, to this day, not realized by those aviators who are hankering for monoplanes.

Experiments by Stringfellow.

The next man to test an apparatus with superposed surfaces was Mr. Stringfellow, who, becoming much impressed with Mr. Wenham's proposal, produced a largish model at the exhibition of the Aeronautical Society in 1868. It consisted of three superposed surfaces aggregating 28 square feet and a tail of 8 square feet more. The weight was under 12 pounds and it was driven by a central propeller actuated by a steam engine overestimated at one-third of a horsepower. It ran suspended to a wire on its trials but failed of free flight, in consequence of defective equilibrium. This apparatus has since been rebuilt and is now in the National Museum of the Smithsonian Institution at Washington. Linfield's Unsuccessful Efforts.

In 1878 Mr. Linfield tested an apparatus in England consisting of a cigar-shaped car, to which was attached on each side frames five feet square, containing each twenty-five superposed planes of stretched and varnished linen eighteen inches wide, and only two inches apart, thus reminding one of a Spanish donkey with panniers. The whole weighed two hundred and forty pounds. This was tested by being mounted on a flat car behind a locomotive going 40 miles an hour. When towed by a line fifteen feet long the apparatus rose only a little from the car and exhibited such unstable equilibrium that the experiment was not renewed. The lift was only about one-third of what it would have been had the planes been properly spaced, say their full width apart, instead of one-ninth as erroneously devised.

Renard's "Dirigible Parachute."

In 1889 Commandant Renard, the eminent superintendent of the French Aeronautical Department, exhibited at the Paris Exposition of that year, an apparatus experimented with some years before, which he termed a "dirigible parachute." It consisted of an oviform body to which were pivoted two upright slats carrying above the body nine long superposed flat blades spaced about one-third of their width apart. When this apparatus was properly set at an angle to the longitudinal axis of the body and dropped from a balloon, it travelled back against the wind for a considerable distance before alighting. The course could be varied by a rudder. No practical application seems to have been made of this device by the French War Department, but Mr. J. P. Holland, the inventor of the submarine boat which bears his name, proposed in 1893 an arrangement of pivoted framework attached to the body of a flying machine which combines the principle of Commandant Renard with the curved blades experimented with by Mr. Phillips, now to be noticed, with the addition of lifting screws inserted among the blades.

Phillips Fails on Stability Problem.

In 1893 Mr. Horatio Phillips, of England, after some very interesting experiments with various wing sections, from which he deduced conclusions as to the shape of maximum lift, tested an apparatus resembling a Venetian blind which consisted of fifty wooden slats of peculiar shape, 22 feet long, one and a half inches wide, and two inches apart, set in ten vertical upright boards. All this was carried upon a body provided with three wheels. It weighed 420 pounds and was driven at 40 miles an hour on a wooden sidewalk by a steam engine of nine horsepower which actuated a two-bladed screw. The lift was satisfactory, being perhaps 70 pounds per horsepower, but the equilibrium was quite bad and the experiments were discontinued. They were taken up again in 1904 with a similar apparatus large enough to carry a passenger, but the longitudinal equilibrium was found to be defective. Then in 1907 a new machine was tested, in which four sets of frames, carrying similar sets of slat "sustainers" were inserted, and with this arrangement the longitudinal stability was found to be very satisfactory. The whole apparatus, with the operator, weighed 650 pounds. It flew about 200 yards when driven by a motor of 20 to 22 h.p. at 30 miles an hour, thus exhibiting a lift of about 32 pounds per h.p., while it will be remembered that the aeroplane of Wright Brothers exhibits a lifting capacity of 50 pounds to the h.p.

Hargrave's Kite Experiments.

After experimenting with very many models and building no less than eighteen monoplane flying model machines, actuated by rubber, by compressed air and by steam, Mr. Lawrence Hargrave, of Sydney, New South Wales, invented the cellular kite which bears his name and made it known in a paper contributed to the Chicago Conference on Aerial Navigation in 1893, describing several varieties. The modern construction is well known, and consists of two cells, each of superposed surfaces with vertical side fins, placed one behind the other and connected by a rod or frame. This flies with great steadiness without a tail. Mr. Hargrave's idea was to use a team of these kites, below which he proposed to suspend a motor and propeller from which a line would be carried to an anchor in the ground. Then by actuating the propeller the whole apparatus would move forward, pick up the anchor and fly away. He said: "The next step is clear enough, namely, that a flying machine with acres of surface can be safely got under way or anchored and hauled to the ground by means of the string of kites."

The first tentative experiments did not result well and emphasized the necessity for a light motor, so that Mr. Hargrave has since been engaged in developing one, not having convenient access to those which have been produced by the automobile designers and builders.

Experiments With Glider Model.

And here a curious reminiscence may be indulged in. In 1888 the present writer experimented with a two-cell gliding model, precisely similar to a Hargrave kite, as will be confirmed by Mr. Herring. It was frequently tested by launching from the top of a three-story house and glided downward very steadily in all sorts of breezes, but the angle of descent was much steeper than that of birds, and the weight sustained per square foot was less than with single cells, in consequence of the lesser support afforded by the rear cell, which operated upon air already set in motion downward by the front cell, so nothing more was done with it, for it never occurred to the writer to try it as a kite and he thus missed the distinction which attaches to Hargrave's name.

Sir Hiram Maxim also introduced fore and aft superposed surfaces in his wondrous flying machine of 1893, but he relied chiefly for the lift upon his main large surface and this necessitated so many guys, to prevent distortion, as greatly to increase the head resistance and this, together with the unstable equilibrium, made it evident that the design of the machine would have to be changed.

How Lilienthal Was Killed.

In 1895, Otto Lilienthal, the father of modern aviation, the man to whose method of experimenting almost all present successes are due, after making something like two thousand glides with monoplanes, added a superposed surface to his apparatus and found the control of it much improved. The two surfaces were kept apart by two struts or vertical posts with a few guy wires, but the connecting joints were weak and there was nothing like trussing. This eventually cost his most useful life. Two weeks before that distressing loss to science, Herr Wilhelm Kress, the distinguished and veteran aviator of Vienna, witnessed a number of glides by Lilienthal with his double-decked apparatus. He noticed that it was much wracked and wobbly and wrote to me after the accident: "The connection of the wings and the steering arrangement were very bad and unreliable. I warned Herr Lilienthal very seriously. He promised me that he would soon put it in order, but I fear that he did not attend to it immediately."

In point of fact, Lilienthal had built a new machine, upon a different principle, from which he expected great results, and intended to make but very few more flights with the old apparatus. He unwisely made one too many and, like Pilcher, was the victim of a distorted apparatus. Probably one of the joints of the struts gave way, the upper surface blew back and Lilienthal, who was well forward on the lower surface, was pitched headlong to destruction.

Experiments by the Writer.

In 1896, assisted by Mr. Herring and Mr. Avery, I experimented with several full sized gliding machines, carrying a man. The first was a Lilienthal monoplane which was deemed so cranky that it was discarded after making about one hundred glides, six weeks before Lilienthal's accident. The second was known as the multiple winged machine and finally developed into five pairs of pivoted wings, trussed together at the front and one pair in the rear. It glided at angles of descent of 10 or 11 degrees or of one in five, and this was deemed too steep. Then Mr. Herring and myself made computations to analyze the resistances. We attributed much of them to the five front spars of the wings and on a sheet of cross-barred paper I at once drew the design for a new three-decked machine to be built by Mr. Herring.

Being a builder of bridges, I trussed these surfaces together, in order to obtain strength and stiffness. When tested in gliding flight the lower surface was found too near the ground. It was taken off and the remaining apparatus now consisted of two surfaces connected together by a girder composed of vertical posts and diagonal ties, specifically known as a "Pratt truss." Then Mr. Herring and Mr. Avery together devised and put on an elastic attachment to the tail. This machine proved a success, it being safe and manageable. Over 700 glides were made with it at angles of descent of 8 to 10 degrees, or one in six to one in seven.

First Proposed by Wenham.

The elastic tail attachment and the trussing of the connecting frame of the superposed wings were the only novelties in this machine, for the superposing of the surfaces had first been proposed by Wenham, but in accordance with the popular perception, which bestows all the credit upon the man who adds the last touch making for success to the labors of his predecessors, the machine has since been known by many persons as the "Chanute type" of gliders, much to my personal gratification.

It has since been improved in many ways. Wright Brothers, disregarding the fashion which prevails among birds, have placed the tail in front of their apparatus and called it a front rudder, besides placing the operator in horizontal position instead of upright, as I did; and also providing a method of warping the wings to preserve equilibrium. Farman and Delagrange, under the very able guidance and constructive work of Voisin brothers, then substituted many details, including a box tail for the dart-like tail which I used. This may have increased the resistance, but it adds to the steadiness. Now the tendency in France seems to be to go back to the monoplane.

Monoplane Idea Wrong.

The advocates of the single supporting surface are probably mistaken. It is true that a single surface shows a greater lift per square foot than superposed surfaces for a given speed, but the increased weight due to leverage more than counterbalances this advantage by requiring heavy spars and some guys. I believe that the future aeroplane dynamic flier will consist of superposed surfaces, and, now that it has been found that by imbedding suitably shaped spars in the cloth the head resistance may be much diminished, I see few objections to superposing three, four or even five surfaces properly trussed, and thus obtaining a compact, handy, manageable and comparatively light apparatus. 2

CHAPTER II. THEORY, DEVELOPMENT, AND USE.

While every craft that navigates the air is an airship, all airships are not flying machines. The balloon, for instance, is an airship, but it is not what is known among aviators as a flying machine. This latter term is properly used only in referring to heavier-than-air machines which have no gas-bag lifting devices, and are made to really fly by the application of engine propulsion.

Mechanical Birds.

All successful flying machines—and there are a number of them—are based on bird action. The various designers have studied bird flight and soaring, mastered its technique as devised by Nature, and the modern flying machine is the result. On an exaggerated, enlarged scale the machines which are now navigating the air are nothing more nor less than mechanical birds.

Origin of the Aeroplane.

Octave Chanute, of Chicago, may well be called "the developer of the flying machine." Leaving balloons and various forms of gas-bags out of consideration, other experimenters, notably Langley and Lilienthal, antedated him in attempting the navigation of the air on aeroplanes, or flying machines, but none of them were wholly successful, and it remained for Chanute to demonstrate the practicability of what was then called the gliding machine. This term was adopted because the apparatus was, as the name implies, simply a gliding machine, being without motor propulsion, and intended solely to solve the problem of the best form of construction. The biplane, used by Chanute in 1896, is still the basis of most successful flying machines, the only radical difference being that motors, rudders, etc., have been added.

Character of Chanute's Experiments.

It was the privilege of the author of this book to be Mr. Chanute's guest at Millers, Indiana, in 1896, when, in collaboration with Messrs. Herring and Avery, he was conducting the series of experiments which have since made possible the construction of the modern flying machine which such successful aviators as the Wright brothers and others are now using. It was a wild country, much frequented by eagles, hawks, and similar birds. The enthusiastic trio, Chanute, Herring and Avery, would watch for hours the evolutions of some big bird in the air, agreeing in the end on the verdict, "When we master the principle of that bird's soaring without wing action, we will have come close to solving the problem of the flying machine."

Aeroplanes of various forms were constructed by Mr. Chanute with the assistance of Messrs. Herring and Avery until, at the time of the writer's visit, they had settled upon the biplane, or two-surface machine. Mr. Herring later equipped this with a rudder, and made other additions, but the general idea is still the basis of the Wright, Curtiss, and other machines in which, by the aid of gasolene motors, long flights have been made.

Developments by the Wrights.

In 1900 the Wright brothers, William and Orville, who were then in the bicycle business in Dayton, Ohio, became interested in Chanute's experiments and communicated with him. The result was that the Wrights took up Chanute's ideas and developed them further, making many additions of their own, one of which was the placing of a rudder in front, and the location of the operator horizontally on the machine, thus diminishing by four-fifths the wind resistance of the man's body. For three years the Wrights experimented with the glider before venturing to add a motor, which was not done until they had thoroughly mastered the control of their movements in the air.

Limits of the Flying Machine.

In the opinion of competent experts it is idle to look for a commercial future for the flying machine. There is, and always will be, a limit to its carrying capacity which will prohibit its employment for passenger or freight purposes in a wholesale or general way. There are some, of course, who will argue that because a machine will carry two people another may be constructed that will carry a dozen, but those who make this contention do not understand the theory of weight sustentation in the air; or that the greater the load the greater must be the lifting power (motors and plane surface), and that there is a limit to these—as will be explained later on—beyond which the aviator cannot go.

Some Practical Uses.

At the same time there are fields in which the flying machine may be used to great advantage. These are:

Sports—Flying machine races or flights will always be popular by reason of the element of danger. It is a strange, but nevertheless a true proposition, that it is this element which adds zest to all sporting events.

Scientific—For exploration of otherwise inaccessible regions such as deserts, mountain tops, etc.

Reconnoitering—In time of war flying machines may be used to advantage to spy out an enemy's encampment, ascertain its defenses, etc.

CHAPTER III. MECHANICAL BIRD ACTION

In order to understand the theory of the modern flying machine one must also understand bird action and wind action. In this connection the following simple experiment will be of interest:

Take a circular-shaped bit of cardboard, like the lid of a hat box, and remove the bent-over portion so as to have a perfectly flat surface with a clean, sharp edge. Holding the cardboard at arm's length, withdraw your hand, leaving the cardboard without support. What is the result? The cardboard, being heavier than air, and having nothing to sustain it, will fall to the ground. Pick it up and throw it, with considerable force, against the wind edgewise. What happens? Instead of falling to the ground, the cardboard sails along on the wind, remaining afloat so long as it is in motion. It seeks the ground, by gravity, only as the motion ceases, and then by easy stages, instead of dropping abruptly as in the first instance.

Here we have a homely, but accurate illustration of the action of the flying machine. The motor does for the latter what the force of your arm does for the cardboard—imparts a motion which keeps it afloat. The only real difference is that the motion given by the motor is continuous and much more powerful than that given by your arm. The action of the latter is limited and the end of its propulsive force is reached within a second or two after it is exerted, while the action of the motor is prolonged.

Another Simple Illustration.

Another simple means of illustrating the principle of flying machine operation, so far as sustentation and the elevation and depression of the planes is concerned, is explained in the accompanying diagram.

A is a piece of cardboard about 2 by 3 inches in size. B is a piece of paper of the same size pasted to one edge of A. If you bend the paper to a curve, with convex side up and blow across it as shown in Figure C, the paper will rise instead of being depressed. The dotted lines show that the air is passing over the top of the curved paper and yet, no matter how hard you may blow, the effect will be to elevate the paper, despite the fact that the air is passing over, instead of under the curved surface.

In Figure D we have an opposite effect. Here the paper is in a curve exactly the reverse of that shown in Figure C, bringing the concave side up. Now if you will again blow across the surface of the card the action of the paper will be downward—it will be impossible to make it rise. The harder you blow the greater will be the downward movement.

Principle In General Use.

This principle is taken advantage of in the construction of all successful flying machines. Makers of monoplanes and biplanes alike adhere to curved bodies, with the concave surface facing downward. Straight planes were tried for a time, but found greatly lacking in the power of sustentation. By curving the planes, and placing the concave surface downward, a sort of inverted bowl is formed in which the air gathers and exerts a buoyant effect. Just what the ratio of the curve should be is a matter of contention. In some instances one inch to the foot is found to be satisfactory; in others this is doubled, and there are a few cases in which a curve of as much as 3 inches to the foot has been used.

Right here it might be well to explain that the word "plane" applied to flying machines of modern construction is in reality a misnomer. Plane indicates a flat, level surface. As most successful flying machines have curved supporting surfaces it is clearly wrong to speak of "planes," or "aeroplanes." Usage, however, has made the terms convenient and, as they are generally accepted and understood by the public, they are used in like manner in this volume.

Getting Under Headway.

A bird, on first rising from the ground, or beginning its flight from a tree, will flap its wings to get under headway. Here again we have another illustration of the manner in which a flying machine gets under headway—the motor imparts the force necessary to put the machine into the air, but right here the similarity ceases. If the machine is to be kept afloat the motor must be kept moving. A flying machine will not sustain itself; it will not remain suspended in the air unless it is under headway. This is because it is heavier than air, and gravity draws it to the ground.

Puzzle in Bird Soaring.

But a bird, which is also heavier than air, will remain suspended, in a calm, will even soar and move in a circle, without apparent movement of its wings. This is explained on the theory that there are generally vertical columns of air in circulation strong enough to sustain a bird, but much too weak to exert any lifting power on a flying machine, It is easy to understand how a bird can remain suspended when the wind is in action, but its suspension in a seeming dead calm was a puzzle to scientists until Mr. Chanute advanced the proposition of vertical columns of air.

Modeled Closely After Birds.

So far as possible, builders of flying machines have taken what may be called "the architecture" of birds as a model. This is readily noticeable in the form of construction. When a bird is in motion its wings (except when flapping) are extended in a straight line at right angles to its body. This brings a sharp, thin edge against the air, offering the least possible surface for resistance, while at the same time a broad surface for support is afforded by the flat, under side of the wings. Identically the same thing is done in the construction of the flying machine.

Note, for instance, the marked similarity in form as shown in the illustration in Chapter II. Here A is the bird, and B the general outline of the machine. The thin edge of the plane in the latter is almost a duplicate of that formed by the outstretched wings of the bird, while the rudder plane in the rear serves the same purpose as the bird's tail.

CHAPTER IV. VARIOUS FORMS OF FLYING MACHINES.

There are three distinct and radically different forms of flying machines. These are:

Aeroplanes, helicopters and ornithopers.

Of these the aeroplane takes precedence and is used almost exclusively by successful aviators, the helicopters and ornithopers having been tried and found lacking in some vital features, while at the same time in some respects the helicopter has advantages not found in the aeroplane.

What the Helicopter Is.

The helicopter gets its name from being fitted with vertical propellers or helices (see illustration) by the action of which the machine is raised directly from the ground into the air. This does away with the necessity for getting the machine under a gliding headway before it floats, as is the case with the aeroplane, and consequently the helicopter can be handled in a much smaller space than is required for an aeroplane. This, in many instances, is an important advantage, but it is the only one the helicopter possesses, and is more than overcome by its drawbacks. The most serious of these is that the helicopter is deficient in sustaining capacity, and requires too much motive power.

Form of the Ornithopter.

The ornithopter has hinged planes which work like the wings of a bird. At first thought this would seem to be the correct principle, and most of the early experimenters conducted their operations on this line. It is now generally understood, however, that the bird in soaring is in reality an aeroplane, its extended wings serving to sustain, as well as propel, the body. At any rate the ornithoper has not been successful in aviation, and has been interesting mainly as an ingenious toy. Attempts to construct it on a scale that would permit of its use by man in actual aerial flights have been far from encouraging.

Three Kinds of Aeroplanes.

There are three forms of aeroplanes, with all of which more or less success has been attained. These are:

The monoplane, a one-surfaced plane, like that used by Bleriot.

The biplane, a two-surfaced plane, now used by the Wrights, Curtiss, Farman, and others.

The triplane, a three-surfaced plane This form is but little used, its only prominent advocate at present being Elle Lavimer, a Danish experimenter, who has not thus far accomplished much.

Whatever of real success has been accomplished in aviation may be credited to the monoplane and biplane, with the balance in favor of the latter. The monoplane is the more simple in construction and, where weight-sustaining capacity is not a prime requisite, may probably be found the most convenient. This opinion is based on the fact that the smaller the surface of the plane the less will be the resistance offered to the air, and the greater will be the speed at which the machine may be moved. On the other hand, the biplane has a much greater plane surface (double that of a monoplane of the same size) and consequently much greater weight-carrying capacity.

Differences in Biplanes.

While all biplanes are of the same general construction so far as the main planes are concerned, each aviator has his own ideas as to the "rigging."

Wright, for instance, places a double horizontal rudder in front, with a vertical rudder in the rear. There are no partitions between the main planes, and the bicycle wheels used on other forms are replaced by skids.

Voisin, on the contrary, divides the main planes with vertical partitions to increase stability in turning; uses a single-plane horizontal rudder in front, and a big box-tail with vertical rudder at the rear; also the bicycle wheels.

Curtiss attaches horizontal stabilizing surfaces to the upper plane; has a double horizontal rudder in front, with a vertical rudder and horizontal stabilizing surfaces in rear. Also the bicycle wheel alighting gear.

CHAPTER V. CONSTRUCTING A GLIDING MACHINE.

First decide upon the kind of a machine you want—monoplane, biplane, or triplane. For a novice the biplane will, as a rule, be found the most satisfactory as it is more compact and therefore the more easily handled. This will be easily understood when we realize that the surface of a flying machine should be laid out in proportion to the amount of weight it will have to sustain. The generally accepted rule is that 152 square feet of surface will sustain the weight of an average-sized man, say 170 pounds. Now it follows that if these 152 square feet of surface are used in one plane, as in the monoplane, the length and width of this plane must be greater than if the same amount of surface is secured by using two planes—the biplane. This results in the biplane being more compact and therefore more readily manipulated than the monoplane, which is an important item for a novice.

Glider the Basis of Success.

Flying machines without motors are called gliders. In making a flying machine you first construct the glider. If you use it in this form it remains a glider. If you install a motor it becomes a flying machine. You must have a good glider as the basis of a successful flying machine.

It will be well for the novice, the man who has never had any experience as an aviator, to begin with a glider and master its construction and operation before he essays the more pretentious task of handling a fully-equipped flying machine. In fact, it is essential that he should do so.

Plans for Handy Glider.

A glider with a spread (advancing edge) of 20 feet, and a breadth or depth of 4 feet, will be about right to begin with. Two planes of this size will give the 152 square yards of surface necessary to sustain a man's weight. Remember that in referring to flying machine measurements "spread" takes the place of what would ordinarily be called "length," and invariably applies to the long or advancing edge of the machine which cuts into the air. Thus, a glider is spoken of as being 20 feet spread, and 4 feet in depth. So far as mastering the control of the machine is concerned, learning to balance one's self in the air, guiding the machine in any desired direction by changing the position of the body, etc., all this may be learned just as readily, and perhaps more so, with a 20-foot glider than with a larger apparatus.

Kind of Material Required.

There are three all-important features in flying machine construction, viz.: lightness, strength and extreme rigidity. Spruce is the wood generally used for glider frames. Oak, ash and hickory are all stronger, but they are also considerably heavier, and where the saving of weight is essential, the difference is largely in favor of spruce. This will be seen in the following table:

Wood	Weight per cubic ft. in lbs.	Tensile Strength lbs. per sq. in.	Compressive Strength lbs. per sq in.
Hickory	53	12,000	8,500
Oak	50	12,000	9,000
Ash	38	12,000	6,000
Walnut	38	8,000	6,000
Spruce	25	8,000	5,000
Pine	25	5,000	4,500

Considering the marked saving in weight spruce has a greater percentage of tensile strength than any of the other woods. It is also easier to find in long, straight-grained pieces free from knots, and it is this kind only that should be used in flying machine construction.

You will next need some spools or hanks of No. 6 linen shoe thread, metal sockets, a supply of strong piano wire, a quantity of closely-woven silk or cotton cloth, glue, turnbuckles, varnish, etc.

Names of the Various Parts.

The long strips, four in number, which form the front and rear edges of the upper and lower frames, are called the horizontal beams. These are each 20 feet in length. These horizontal beams are connected by upright strips, 4 feet long, called stanchions. There are usually 12 of these, six on the front edge, and six on the rear. They serve to hold the upper plane away from the lower one. Next comes the ribs. These are 4 feet in length (projecting for a foot over the rear beam), and while intended principally as a support to the cloth covering of the planes, also tend to hold the frame together in a horizontal position just as the stanchions do in the vertical. There are forty-one of these ribs, twenty-one on the upper and twenty on the lower plane. Then come the struts, the main pieces which join the horizontal beams. All of these parts are shown in the illustrations, reference to which will make the meaning of the various names clear.

Quantity and Cost of Material.

For the horizontal beams four pieces of spruce, 20 feet long, 1 1/2 inches wide and 3/4 inch thick are necessary. These pieces must be straight-grain, and absolutely free from knots. If it is impossible to obtain clear pieces of this length, shorter ones may be spliced, but this is not advised as it adds materially to the weight. The twelve stanchions should be 4 feet long and 7/8 inch in diameter and rounded in form so as to offer as little resistance as possible to the wind. The struts, there are twelve of them, are 3 feet long by 11/4 x 1/2 inch. For a 20-foot biplane about 20 yards of stout silk or unbleached muslin, of standard one yard width, will be needed. The forty-one ribs are each 4 feet long, and 1/2 inch square. A roll of No. 12 piano wire, twenty-four sockets, a package of small copper tacks, a pot of glue, and similar accessories will be required. The entire cost of this material should not exceed $20. The wood and cloth will be the two largest items, and these should not cost more than $10. This leaves $10 for the varnish, wire, tacks, glue, and other incidentals. This estimate is made for cost of materials only, it being taken for granted that the experimenter will construct his own glider. Should the services of a carpenter be required the total cost will probably approximate $60 or $70.

Application of the Rudders.

The figures given also include the expense of rudders, but the details of these have not been included as the glider is really complete without them. Some of the best flights the writer ever saw were made by Mr. A. M. Herring in a glider without a rudder, and yet there can be no doubt that a rudder, properly proportioned and placed, especially a rear rudder, is of great value to the aviator as it keeps the machine with its head to the wind, which is the only safe position for a novice. For initial educational purposes, however, a rudder is not essential as the glides will, or should, be made on level ground, in moderate, steady wind currents, and at a modest elevation. The addition of a rudder, therefore, may well be left until the aviator has become reasonably expert in the management of his machine.

Putting the Machine Together.

Having obtained the necessary material, the first move is to have the rib pieces steamed and curved. This curve may be slight, about 2 inches for the 4 feet. While this is being done the other parts should be carefully rounded so the square edges will be taken off. This may be done with sand paper. Next apply a coat of shellac, and when dry rub it down thoroughly with fine sand paper. When the ribs are curved treat them in the same way.

Lay two of the long horizontal frame pieces on the floor 3 feet apart. Between these place six of the strut pieces. Put one at each end, and each 4 1/2 feet put another, leaving a 2-foot space in the center. This will give you four struts 4 1/2 feet apart, and two in the center 2 feet apart, as shown in the illustration. This makes five rectangles. Be sure that the points of contact are perfect, and that the struts are exactly at right angles with the horizontal frames. This is a most important feature because if your frame "skews" or twists you cannot keep it straight in the air. Now glue the ends of the struts to the frame pieces, using plenty of glue, and nail on strips that will hold the frame in place while the glue is drying. The next day lash the joints together firmly with the shoe thread, winding it as you would to mend a broken gun stock, and over each layer put a coating of glue. This done, the other frame pieces and struts may be treated in the same way, and you will thus get the foundations for the two planes.

Another Way of Placing Struts.

In the machines built for professional use a stronger and more certain form of construction is desired. This is secured by the placing the struts for the lower plane under the frame piece, and those for the upper plane over it, allowing them in each instance to come out flush with the outer edges of the frame pieces. They are then securely fastened with a tie plate or clamp which passes over the end of the strut and is bound firmly against the surface of the frame piece by the eye bolts of the stanchion sockets.

Placing the Rib Pieces.

Take one of the frames and place on it the ribs, with the arched side up, letting one end of the ribs come flush with the front edge of the forward frame, and the other end projecting about a foot beyond the rear frame. The manner of fastening the ribs to the frame pieces is optional. In some cases they are lashed with shoe thread, and in others clamped with a metal clamp fastened with 1/2-inch wood screws. Where clamps and screws are used care should be taken to make slight holes in the wood with an awl before starting the screws so as to lessen any tendency to split the wood. On the top frame, twenty-one ribs placed one foot apart will be required. On the lower frame, because of the opening left for the operator's body, you will need only twenty.

Joining the Two Frames.

The two frames must now be joined together. For this you will need twenty-four aluminum or iron sockets which may be purchased at a foundry or hardware shop. These sockets, as the name implies, provide a receptacle in which the end of a stanchion is firmly held, and have flanges with holes for eye-bolts which hold them firmly to the frame pieces, and also serve to hold the guy wires. In addition to these eye-bolt holes there are two others through which screws are fastened into the frame pieces. On the front frame piece of the bottom plane place six sockets, beginning at the end of the frame, and locating them exactly opposite the struts. Screw the sockets into position with wood screws, and then put the eye-bolts in place. Repeat the operation on the rear frame. Next put the sockets for the upper plane frame in place.

You are now ready to bring the two planes together. Begin by inserting the stanchions in the sockets in the lower plane. The ends may need a little rubbing with sandpaper to get them into the sockets, but care must be taken to have them fit snugly. When all the stanchions are in place on the lower plane, lift the upper plane into position, and fit the sockets over the upper ends of the stanchions.

Trussing with Guy Wires.

The next move is to "tie" the frame together rigidly by the aid of guy wires. This is where the No. 12 piano wire comes in. Each rectangle formed by the struts and stanchions with the exception of the small center one, is to be wired separately as shown in the illustration. At each of the eight corners forming the rectangle the ring of one of the eye-bolts will be found. There are two ways of doing this "tieing," or trussing. One is to run the wires diagonally from eye-bolt to eye-bolt, depending upon main strength to pull them taut enough, and then twist the ends so as to hold. The other is to first make a loop of wire at each eye-bolt, and connect these loops to the main wires with turn-buckles. This latter method is the best, as it admits of the tension being regulated by simply turning the buckle so as to draw the ends of the wire closer together. A glance at the illustration will make this plain, and also show how the wires are to be placed. The proper degree of tension may be determined in the following manner:

After the frame is wired place each end on a saw-horse so as to lift the entire frame clear of the work-shop floor. Get under it, in the center rectangle and, grasping the center struts, one in each hand, put your entire weight on the structure. If it is properly put together it will remain rigid and unyielding. Should it sag ever so slightly the tension of the wires must be increased until any tendency to sag, no matter how slight it may be, is overcome.

Putting on the Cloth.

We are now ready to put on the cloth covering which holds the air and makes the machine buoyant. The kind of material employed is of small account so long as it is light, strong, and wind-proof, or nearly so. Some aviators use what is called rubberized silk, others prefer balloon cloth. Ordinary muslin of good quality, treated with a coat of light varnish after it is in place, will answer all the purposes of the amateur.

Cut the cloth into strips a little over 4 feet in length. As you have 20 feet in width to cover, and the cloth is one yard wide, you will need seven strips for each plane, so as to allow for laps, etc. This will give you fourteen strips. Glue the end of each strip around the front horizontal beams of the planes, and draw each strip back, over the ribs, tacking the edges to the ribs as you go along, with small copper or brass tacks. In doing this keep the cloth smooth and stretched tight. Tacks should also be used in addition to the glue, to hold the cloth to the horizontal beams.

Next, give the cloth a coat of varnish on the clear, or upper side, and when this is dry your glider will be ready for use.

Reinforcing the Cloth.

While not absolutely necessary for amateur purposes, reinforcement of the cloth, so as to avoid any tendency to split or tear out from wind-pressure, is desirable. One way of doing this is to tack narrow strips of some heavier material, like felt, over the cloth where it laps on the ribs. Another is to sew slips or pockets in the cloth itself and let the ribs run through them. Still another method is to sew 2-inch strips (of the same material as the cover) on the cloth, placing them about one yard apart, but having them come in the center of each piece of covering, and not on the laps where the various pieces are joined.

Use of Armpieces.

Should armpieces be desired, aside from those afforded by the center struts, take two pieces of spruce, 3 feet long, by 1 x 1 3/4 inches, and bolt them to the front and rear beams of the lower plane about 14 inches apart. These will be more comfortable than using the struts, as the operator will not have to spread his arms so much. In using the struts the operator, as a rule, takes hold of them with his hands, while with the armpieces, as the name implies, he places his arms over them, one of the strips coming under each armpit.

Frequently somebody asks why the ribs should be curved. The answer is easy. The curvature tends to direct the air downward toward the rear and, as the air is thus forced downward, there is more or less of an impact which assists in propelling the aeroplane upwards.

CHAPTER VI. LEARNING TO FLY.

Don't be too ambitious at the start. Go slow, and avoid unnecessary risks. At its best there is an element of danger in aviation which cannot be entirely eliminated, but it may be greatly reduced and minimized by the use of common sense.

Theoretically, the proper way to begin a glide is from the top of an incline, facing against the wind, so that the machine will soar until the attraction of gravitation draws it gradually to the ground. This is the manner in which experienced aviators operate, but it must be kept in mind that these men are experts. They understand air currents, know how to control the action and direction of their machines by shifting the position of their bodies, and by so doing avoid accidents which would be unavoidable by a novice.

Begin on Level Ground.

Make your first flights on level ground, having a couple of men to assist you in getting the apparatus under headway. Take your position in the center rectangle, back far enough to give the forward edges of the glider an inclination to tilt upward very slightly. Now start and run forward at a moderately rapid gait, one man at each end of the glider assisting you. As the glider cuts into the air the wind will catch under the uplifted edges of the curved planes, and buoy it up so that it will rise in the air and take you with it. This rise will not be great, just enough to keep you well clear of the ground. Now project your legs a little to the front so as to shift the center of gravity a trifle and bring the edges of the glider on an exact level with the atmosphere. This, with the momentum acquired in the start, will keep the machine moving forward for some distance.

Effect of Body Movements.

When the weight of the body is slightly back of the center of gravity the edges of the advancing planes are tilted slightly upward. The glider in this position acts as a scoop, taking in the air which, in turn, lifts it off the ground. When a certain altitude is reached—this varies with the force of the wind—the tendency to a forward movement is lost and the glider comes to the ground. It is to prolong the forward movement as much as possible that the operator shifts the center of gravity slightly, bringing the apparatus on an even keel as it were by lowering the advancing edges. This done, so long as there is momentum enough to keep the glider moving, it will remain afloat.

If you shift your body well forward it will bring the front edges of the glider down, and elevate the rear ones. In this way the air will be "spilled" out at the rear, and, having lost the air support or buoyancy, the glider comes down to the ground. A few flights will make any ordinary man proficient in the control of his apparatus by his body movements, not only as concerns the elevating and depressing of the advancing edges, but also actual steering. You will quickly learn, for instance, that, as the shifting of the bodily weight backwards and forwards affects the upward and downward trend of the planes, so a movement sideways—to the left or the right—affects the direction in which the glider travels.

Ascends at an Angle.

In ascending, the glider and flying machine, like the bird, makes an angular, not a vertical flight. Just what this angle of ascension may be is difficult to determine. It is probable and in fact altogether likely, that it varies with the force of the wind, weight of the rising body, power of propulsion, etc. This, in the language of physicists, is the angle of inclination, and, as a general thing, under normal conditions (still air) should be put down as about one in ten, or 5 3/4 degrees. This would be an ideal condition, but it has not, as vet been reached. The force of the wind affects the angle considerably, as does also the weight and velocity of the apparatus. In general practice the angle varies from 23 to 45 degrees. At more than 45 degrees the supporting effort is overcome by the resistance to forward motion.

Increasing the speed or propulsive force, tends to lessen the angle at which the machine may be successfully operated because it reduces the wind pressure. Most of the modern flying machines are operated at an angle of 23 degrees, or less.

Maintaining an Equilibrium.

Stable equilibrium is one of the main essentials to successful flight, and this cannot be preserved in an uncertain, gusty wind, especially by an amateur. The novice should not attempt a glide unless the conditions are just right. These conditions are: A clear, level space, without obstructions, such as trees, etc., and a steady wind of not exceeding twelve miles an hour. Always fly against the wind.

When a reasonable amount of proficiency in the handling of the machine on level ground has been acquired the field of practice may be changed to some gentle slope. In starting from a slope it will be found easier to keep the machine afloat, but the experience at first is likely to be very disconcerting to a man of less than iron nerve. As the glider sails away from the top of the slope the distance between him and the ground increases rapidly until the aviator thinks he is up a hundred miles in the air. If he will keep cool, manipulate his apparatus so as to preserve its equilibrium, and "let nature take its course," he will come down gradually and safely to the ground at a considerable distance from the starting place. This is one advantage of starting from an elevation—your machine will go further.

But, if the aviator becomes "rattled"; if he loses control of his machine, serious results, including a bad fall with risk of death, are almost certain. And yet this practice is just as necessary as the initial lessons on level ground. When judgment is used, and "haste made slowly," there is very little real danger. While experimenting with gliders the Wrights made flights innumerable under all sorts of conditions and never had an accident of any kind.

Effects of Wind Currents.

The larger the machine the more difficult it will be to control its movements in the air, and yet enlargement is absolutely necessary as weight, in the form of motor, rudder, etc., is added.

Air currents near the surface of the ground are diverted by every obstruction unless the wind is blowing hard enough to remove the obstruction entirely. Take, for instance, the case of a tree or shrub, in a moderate wind of from ten to twelve miles an hour. As the wind strikes the tree it divides, part going to one side and part going to the other, while still another part is directed upward and goes over the top of the obstruction. This makes the handling of a glider on an obstructed field difficult and uncertain. To handle a glider successfully the place of operation should be clear and the wind moderate and steady. If it is gusty postpone your flight. In this connection it will be well to understand the velocity of the wind, and what it means as shown in the following table:

Miles per hour Feet per second Pressure per sq. foot 10 14.7 .492 25 36.7 3.075 50 73.3 12.300 100 146.6 49.200

Pressure of wind increases in proportion to the square of the velocity. Thus wind at 10 miles an hour has four times the pressure of wind at 5 miles an hour. The greater this pressure the large and heavier the object which can be raised. Any boy who has had experience in flying kites can testify to this, High winds, however, are almost invariably gusty and uncertain as to direction, and this makes them dangerous for aviators. It is also a self-evident fact that, beyond a certain stage, the harder the wind blows the more difficult it is to make headway against it.

Launching Device for Gliders.

On page 195 will be found a diagram of the various parts of a launcher for gliders, designed and patented by Mr. Octave Chanute. In describing this invention in Aeronautics, Mr. Chanute says:

"In practicing, the track, preferably portable, is generally laid in the direction of the existing wind and the car, preferably a light platform-car, is placed on the track. The truck carrying the winding-drum and its motor is placed to windward a suitable distance—say from two hundred to one thousand feet—and is firmly blocked or anchored in line with the portable track, which is preferably 80 or 100 feet in length. The flying or gliding machine to be launched with its operator is placed on the platform-car at the leeward end of the portable track. The line, which is preferably a flexible combination wire-and-cord cable, is stretched between the winding-drum on the track and detachably secured to the flying or gliding machine, preferably by means of a trip-hoop, or else held in the hand of the operator, so that the operator may readily detach the same from the flying-machine when the desired height is attained."

How Glider Is Started.

"Then upon a signal given by the operator the engineer at the motor puts it into operation, gradually increasing the speed until the line is wound upon the drum at a maximum speed of, say, thirty miles an hour. The operator of the flying-machine, whether he stands upright and carries it on his shoulders, or whether he sits or lies down prone upon it, adjusts the aeroplane or carrying surfaces so that the wind shall strike them on the top and press downward instead of upward until the platform-car under action of the winding-drum and line attains the required speed.

"When the operator judges that his speed is sufficient, and this depends upon the velocity of the wind as well as that of the car moving against the wind, he quickly causes the front of the flying-machine to tip upward, so that the relative wind striking on the under side of the planes or carrying surfaces shall lift the flying machine into the air. It then ascends like a kite to such height as may be desired by the operator, who then trips the hook and releases the line from the machine."

What the Operator Does.

"The operator being now free in the air has a certain initial velocity imparted by the winding-drum and line and also a potential energy corresponding to his height above the ground. If the flying or gliding machine is provided with a motor, he can utilize that in his further flight, and if it is a simple gliding machine without motor he can make a descending flight through the air to such distance as corresponds to the velocity acquired and the height gained, steering meanwhile by the devices provided for that purpose.

"The simplest operation or maneuver is to continue the flight straight ahead against the wind; but it is possible to vary this course to the right or left, or even to return in downward flight with the wind to the vicinity of the starting-point. Upon nearing the ground the operator tips upward his carrying-surfaces and stops his headway upon the cushion of increased air resistance so caused. The operator is in no way permanently fastened to his machine, and the machine and the operator simply rest upon the light platform-car, so that the operator is free to rise with the machine from the car whenever the required initial velocity is attained.

Motor For the Launcher.

"The motor may be of any suitable kind or construction, but is preferably an electric or gasolene motor. The winding-drum is furnished with any suitable or customary reversing-guide to cause the line to wind smoothly and evenly upon the drum. The line is preferably a cable composed of flexible wire and having a cotton or other cord core to increase its flexibility. The line extends from the drum to the flying or gliding machine. Its free end may, if desired, be grasped and held by the operator until the flying-machine ascends to the desired height, when by simply letting go of the line the operator may continue his flight free. The line, however, is preferably connected to the flying or gliding machine directly by a trip-hook having a handle or trip lever within reach of the operator, so that when he ascends to the required height he may readily detach the line from the flying or gliding machine."

CHAPTER VII. PUTTING ON THE RUDDER.

Gliders as a rule have only one rudder, and this is in the rear. It tends to keep the apparatus with its head to the wind. Unlike the rudder on a boat it is fixed and immovable. The real motor-propelled flying machine, generally has both front and rear rudders manipulated by wire cables at the will of the operator.

Allowing that the amateur has become reasonably expert in the manipulation of the glider he should, before constructing an actual flying machine, equip his glider with a rudder.

Cross Pieces for Rudder Beam.

To do this he should begin by putting in a cross piece, 2 feet long by 1/4 x 3/4 inches between the center struts, in the lower plane. This may be fastened to the struts with bolts or braces. The former method is preferable. On this cross piece, and on the rear frame of the plane itself, the rudder beam is clamped and bolted. This rudder beam is 8 feet 11 inches long. Having put these in place duplicate them in exactly the same manner and dimensions from the upper frame The cross pieces on which the ends of the rudder beams are clamped should be placed about one foot in advance of the rear frame beam.

The Rudder Itself.

The next step is to construct the rudder itself. This consists of two sections, one horizontal, the other vertical. The latter keeps the aeroplane headed into the wind, while the former keeps it steady—preserves the equilibrium.

The rudder beams form the top and bottom frames of the vertical rudder. To these are bolted and clamped two upright pieces, 3 feet, 10 inches in length, and 3/4 inch in cross section. These latter pieces are placed about two feet apart. This completes the framework of the vertical rudder. See next page (59).

For the horizontal rudder you will require two strips 6 feet long, and four 2 feet long. Find the exact center of the upright pieces on the vertical rudder, and at this spot fasten with bolts the long pieces of the horizontal, placing them on the outside of the vertical strips. Next join the ends of the horizontal strips with the 2-foot pieces, using small screws and corner braces. This done you will have two of the 2-foot pieces left. These go in the center of the horizontal frame, "straddling" the vertical strips, as shown in the illustration.

The framework is to be covered with cloth in the same manner as the planes. For this about ten yards will be needed.

Strengthening the Rudder.

To ensure rigidity the rudder must be stayed with guy wires. For this purpose the No. 12 piano wire is the best. Begin by running two of these wires from the top eye-bolts of stanchions 3 and 4, page 37, to rudder beam where it joins the rudder planes, fastening them at the bottom. Then run two wires from the top of the rudder beam at the same point, to the bottom eye-bolts of the same stanchions. This will give you four diagonal wires reaching from the rudder beam to the top and bottom planes of the glider. Now, from the outer ends of the rudder frame run four similar diagonal wires to the end of the rudder beam where it rests on the cross piece. You will then have eight truss wires strengthening the connection of the rudder to the main body of the glider.

The framework of the rudder planes is then to be braced in the same way, which will take eight more wires, four for each rudder plane. All the wires are to be connected at one end with turn-buckles so the tension may be regulated as desired.

In forming the rudder frame it will be well to mortise the corners, tack them together with small nails, and then put in a corner brace in the inside of each joint. In doing this bear in mind that the material to be thus fastened is light, and consequently the lightest of nails, screws, bolts and corner pieces, etc., is necessary.

CHAPTER VIII. THE REAL FLYING MACHINE.

We will now assume that you have become proficient enough to warrant an attempt at the construction of a real flying machine — one that will not only remain suspended in the air at the will of the operator, but make respectable progress in whatever direction he may desire to go. The glider, it must be remembered, is not steerable, except to a limited extent, and moves only in one direction — against the wind. Besides this its power of flotation — suspension in the air — is circumscribed.

Larger Surface Area Required.

The real flying machine is the glider enlarged, and equipped with motor and propeller. The first thing to do is to decide upon the size required. While a glider of 20 foot spread is large enough to sustain a man it could not under any possible conditions, be made to rise with the weight of the motor, propeller and similar equipment added. As the load is increased so must the surface area of the planes be increased. Just what this increase in surface area should be is problematical as experienced aviators disagree, but as a general proposition it may be placed at from three to four times the area of a 20-foot glider. 3

Some Practical Examples.

The Wrights used a biplane 41 feet in spread, and 6 1/2 ft. deep. This, for the two planes, gives a total surface area of 538 square feet, inclusive of auxiliary planes. This sustains the engine equipment, operator, etc., a total weight officially announced at 1,070 pounds. It shows a lifting capacity of about two pounds to the square foot of plane surface, as against a lifting capacity of about 1/2 pound per square foot of plane surface for the 20-foot glider. This same Wright machine is also reported to have made a successful flight, carrying a total load of 1,100 pounds, which would be over two pounds for each square foot of surface area, which, with auxiliary planes, is 538 square feet.

To attain the same results in a monoplane, the single surface would have to be 60 feet in spread and 9 feet deep. But, while this is the mathematical rule, Bleriot has demonstrated that it does not always hold good. On his record-breaking trip across the English channel, July 25th, 1909, the Frenchman was carried in a monoplane 24 1/2 feet in spread, and with a total sustaining surface of 150 1/2 square feet. The total weight of the outfit, including machine, operator and fuel sufficient for a three-hour run, was only 660 pounds. With an engine of (nominally) 25 horsepower the distance of 21 miles was covered in 37 minutes.

Which is the Best?

Right here an established mathematical quantity is involved. A small plane surface offers less resistance to the air than a large one and consequently can attain a higher rate of speed. As explained further on in this chapter speed is an important factor in the matter of weight-sustaining capacity. A machine that travels one-third faster than another can get along with one-half the surface area of the latter without affecting the load. See the closing paragraph of this chapter on this point. In theory the construction is also the simplest, but this is not always found to be so in practice. The designing and carrying into execution of plans for an extensive area like that of a monoplane involves great skill and cleverness in getting a framework that will be strong enough to furnish the requisite support without an undue excess of weight. This proposition is greatly simplified in the biplane and, while the speed attained by the latter may not be quite so great as that of the monoplane, it has much larger weight-carrying capacity.

Proper Sizes For Frame.

Allowing that the biplane form is selected the construction may be practically identical with that of the 20-foot glider described in Chapter V., except as to size and elimination of the armpieces. In size the surface planes should be about twice as large as those of the 20-foot glider, viz: 40 feet spread instead of 20, and 6 feet deep instead of 3. The horizontal beams, struts, stanchions, ribs, etc., should also be increased in size proportionately.

While care in the selection of clear, straight-grained timber is important in the glider, it is still more important in the construction of a motor-equipped flying machine as the strain on the various parts will be much greater.

How to Splice Timbers.

It is practically certain that you will have to resort to splicing the horizontal beams as it will be difficult, if not impossible, to find 40-foot pieces of timber totally free from knots and worm holes, and of straight grain.

If splicing is necessary select two good 20-foot pieces, 3 inches wide and 1 1/2 inches thick, and one 10-foot long, of the same thickness and width. Plane off the bottom sides of the 10-foot strip, beginning about two feet back from each end, and taper them so the strip will be about 3/4 inch thick at the extreme ends. Lay the two 20-foot beams end to end, and under the joint thus made place the 10-foot strip, with the planed-off ends downward. The joint of the 20-foot pieces should be directly in the center of the 10-foot piece. Bore ten holes (with a 1/4-inch augur) equi-distant apart through the 20-foot strips and the 10-foot strip under them. Through these holes run 1/4-inch stove bolts with round, beveled heads. In placing these bolts use washers top and bottom, one between the head and the top beam, and the other between the bottom beam and the screw nut which holds the bolt. Screw the nuts down hard so as to bring the two beams tightly together, and you will have a rigid 40-foot beam.

Splicing with Metal Sleeves.

An even better way of making a splice is by tonguing and grooving the ends of the frame pieces and enclosing them in a metal sleeve, but it requires more mechanical skill than the method first named. The operation of tonguing and grooving is especially delicate and calls for extreme nicety of touch in the handling of tools, but if this dexterity is possessed the job will be much more satisfactory than one done with a third timber.

As the frame pieces are generally about 1 1/2 inch in diameter, the tongue and the groove into which the tongue fits must be correspondingly small. Begin by sawing into one side of one of the frame pieces about 4 inches back from the end. Make the cut about 1/2 inch deep. Then turn the piece over and duplicate the cut. Next saw down from the end to these cuts. When the sawed-out parts are removed you will have a "tongue" in the end of the frame timber 4 inches long and 1/2 inch thick. The next move is to saw out a 5/8-inch groove in the end of the frame piece which is to be joined. You will have to use a small chisel to remove the 5/8-inch bit. This will leave a groove into which the tongue will fit easily.

Joining the Two Pieces.

Take a thin metal sleeve—this is merely a hollow tube of aluminum or brass open at each end—8 inches long, and slip it over either the tongued or grooved end of one of the frame timbers. It is well to have the sleeve fit snugly, and this may necessitate a sandpapering of the frame pieces so the sleeve will slip on.

Push the sleeve well back out of the way. Cover the tongue thoroughly with glue, and also put some on the inside of the groove. Use plenty of glue. Now press the tongue into the groove, and keep the ends firmly together until the glue is thoroughly dried. Rub off the joint lightly with sand-paper to remove any of the glue which may have oozed out, and slip the sleeve into place over the joint. Tack the sleeve in position with small copper tacks, and you will have an ideal splice.

The same operation is to be repeated on each of the four frame pieces. Two 20-foot pieces joined in this way will give a substantial frame, but when suitable timber of this kind can not be had, three pieces, each 6 feet 11 inches long, may be used. This would give 20 feet 9 inches, of which 8 inches will be taken up in the two joints, leaving the frame 20 feet 1 inch long.

Installation of Motor.

Next comes the installation of the motor. The kinds and efficiency of the various types are described in the following chapter (IX). All we are interested in at this point is the manner of installation. This varies according to the personal ideas of the aviator. Thus one man puts his motor in the front of his machine, another places it in the center, and still another finds the rear of the frame the best. All get good results, the comparative advantages of which it is difficult to estimate. Where one man, as already explained, flies faster than another, the one beaten from the speed standpoint has an advantage in the matter of carrying weight, etc.

The ideas of various well-known aviators as to the correct placing of motors may be had from the following:

Wrights — In rear of machine and to one side.

Curtiss — Well to rear, about midway between upper and lower planes.

Raich — In rear, above the center.

Brauner-Smith — In exact center of machine.

Van Anden — In center.

Herring-Burgess — Directly behind operator.

Voisin — In rear, and on lower plane.

Bleriot — In front.

R. E. P. — In front.

The One Chief Object.

An even distribution of the load so as to assist in maintaining the equilibrium of the machine, should be the one chief object in deciding upon the location of the motor. It matters little what particular spot is selected so long as the weight does not tend to overbalance the machine, or to "throw it off an even keel." It is just like loading a vessel, an operation in which the expert seeks to so distribute the weight of the cargo as to keep the vessel in a perfectly upright position, and prevent a "list" or leaning to one side. The more evenly the cargo is distributed the more perfect will be the equilibrium of the vessel and the better it can be handled. Sometimes, when not properly stowed, the cargo shifts, and this at once affects the position of the craft. When a ship "lists" to starboard or port a preponderating weight of the cargo has shifted sideways; if bow or stern is unduly depressed it is a sure indication that the cargo has shifted accordingly. In either event the handling of the craft becomes not only difficult, but extremely hazardous. Exactly the same conditions prevail in the handling of a flying machine.

Shape of Machine a Factor.

In placing the motor you must be governed largely by the shape and construction of the flying machine frame. If the bulk of the weight of the machine and auxiliaries is toward the rear, then the natural location for the motor will be well to the front so as to counterbalance the excess in rear weight. In the same way if the preponderance of the weight is forward, then the motor should be placed back of the center.

As the propeller blade is really an integral part of the motor, the latter being useless without it, its placing naturally depends upon the location selected for the motor.

Rudders and Auxiliary Planes.

Here again there is great diversity of opinion among aviators as to size, location and form. The striking difference of ideas in this respect is well illustrated in the choice made by prominent makers as follows:

Voisin—horizontal rudder, with two wing-like planes, in front; box-like longitudinal stability plane in rear, inside of which is a vertical rudder.

Wright—large biplane horizontal rudder in front at considerable distance—about 10 feet—from the main planes; vertical biplane rudder in rear; ends of upper and lower main planes made flexible so they may be moved.

Curtiss—horizontal biplane rudder, with vertical damping plane between the rudder planes about 10 feet in front of main planes; vertical rudder in rear; stabilizing planes at each end of upper main plane.

Bleriot—V-shaped stabilizing fin, projecting from rear of plane, with broad end outward; to the broad end of this fin is hinged a vertical rudder; horizontal biplane rudder, also in rear, under the fin.

These instances show forcefully the wide diversity of opinion existing among experienced aviators as to the best manner of placing the rudders and stabilizing, or auxiliary planes, and make manifest how hopeless would be the task of attempting to select any one form and advise its exclusive use.

Rudder and Auxiliary Construction.

The material used in the construction of the rudders and auxiliary planes is the same as that used in the main planes—spruce for the framework and some kind of rubberized or varnished cloth for the covering. The frames are joined and wired in exactly the same manner as the frames of the main planes, the purpose being to secure the same strength and rigidity. Dimensions of the various parts depend upon the plan adopted and the size of the main plane.

No details as to exact dimensions of these rudders and auxiliary planes are obtainable. The various builders, while willing enough to supply data as to the general measurements, weight, power, etc., of their machines, appear to have overlooked the details of the auxiliary parts, thinking, perhaps, that these were of no particular import to the general public. In the Wright machine, the rear horizontal and front vertical rudders may be set down as being about one-quarter (probably a little less) the size of the main supporting planes.

Arrangement of Alighting Gear.

Most modern machines are equipped with an alighting gear, which not only serves to protect the machine and aviator from shock or injury in touching the ground, but also aids in getting under headway. All the leading makes, with the exception of the Wright, are furnished with a frame carrying from two to five pneumatic rubber-tired bicycle wheels. In the Curtiss and Voisin machines one wheel is placed in front and two in the rear. In the Bleriot and other prominent machines the reverse is the rule—two wheels in front and one in the rear. Farman makes use of five wheels, one in the extreme rear, and four, arranged in pairs, a little to the front of the center of the main lower plane.

In place of wheels the Wright machine is equipped with a skid-like device consisting of two long beams attached to the lower plane by stanchions and curving up far in front, so as to act as supports to the horizontal rudder.

Why Wood Is Favored.

A frequently asked question is: "Why is not aluminum, or some similar metal, substituted for wood." Wood, particularly spruce, is preferred because, weight considered, it is much stronger than aluminum, and this is the lightest of all metals. In this connection the following table will be of interest:

Compressive Weight Tensile Strength Strength per cubic foot per sq. inch per sq. inch Material in lbs. in lbs. in lbs.

Spruce.... 25 8,000 5,000 Aluminum 162 16,000 ...... Brass (sheet) 510 23,000 12,000 Steel (tool) 490 100,000 40,000 Copper (sheet) 548 30,000 40,000

As extreme lightness, combined with strength, especially tensile strength, is the great essential in flying-machine construction, it can be readily seen that the use of metal, even aluminum, for the framework, is prohibited by its weight. While aluminum has double the strength of spruce wood it is vastly heavier, and thus the advantage it has in strength is overbalanced many times by its weight. The specific gravity of aluminum is 2.50; that of spruce is only 0.403.

Things to Be Considered.

In laying out plans for a flying machine there are five important points which should be settled upon before the actual work of construction is started. These are:

First—Approximate weight of the machine when finished and equipped.

Second—Area of the supporting surface required.

Third—Amount of power that will be necessary to secure the desired speed and lifting capacity.

Fourth—Exact dimensions of the main framework and of the auxiliary parts.

Fifth—Size, speed and character of the propeller.

In deciding upon these it will be well to take into consideration the experience of expert aviators regarding these features as given elsewhere. (See Chapter X.)

Estimating the Weights Involved.

In fixing upon the probable approximate weight in advance of construction much, of course, must be assumed. This means that it will be a matter of advance estimating. If a two-passenger machine is to be built we will start by assuming the maximum combined weight of the two people to be 350 pounds. Most of the professional aviators are lighter than this. Taking the medium between the weights of the Curtiss and Wright machines we have a net average of 850 pounds for the framework, motor, propeller, etc. This, with the two passengers, amounts to 1,190 pounds. As the machines quoted are in successful operation it will be reasonable to assume that this will be a safe basis to operate on.

What the Novice Must Avoid.

This does not mean, however, that it will be safe to follow these weights exactly in construction, but that they will serve merely as a basis to start from. Because an expert can turn out a machine, thoroughly equipped, of 850 pounds weight, it does not follow that a novice can do the same thing. The expert's work is the result of years of experience, and he has learned how to construct frames and motor plants of the utmost lightness and strength.

It will be safer for the novice to assume that he can not duplicate the work of such men as Wright and Curtiss without adding materially to the gross weight of the framework and equipment minus passengers.

How to Distribute the Weight.

Let us take 1,030 pounds as the net weight of the machine as against the same average in the Wright and Curtiss machines. Now comes the question of distributing this weight between the framework, motor, and other equipment. As a general proposition the framework should weigh about twice as much as the complete power plant (this is for amateur work).

The word "framework" indicates not only the wooden frames of the main planes, auxiliary planes, rudders, etc., but the cloth coverings as well—everything in fact except the engine and propeller.

On the basis named the framework would weigh 686 pounds, and the power plant 344. These figures are liberal, and the results desired may be obtained well within them as the novice will learn as he makes progress in the work.

Figuring on Surface Area.

It was Prof. Langley who first brought into prominence in connection with flying machine construction the mathematical principle that the larger the object the smaller may be the relative area of support. As explained in Chapter XIII, there are mechanical limits as to size which it is not practical to exceed, but the main principle remains in effect.

Take two aeroplanes of marked difference in area of surface. The larger will, as a rule, sustain a greater weight in relative proportion to its area than the smaller one, and do the work with less relative horsepower. As a general thing well-constructed machines will average a supporting capacity of one pound for every one-half square foot of surface area. Accepting this as a working rule we find that to sustain a weight of 1,200 pounds—machine and two passengers—we should have 600 square feet of surface.

Distributing the Surface Area.

The largest surfaces now in use are those of the Wright, Voisin and Antoinette machines—538 square feet in each. The actual sustaining power of these machines, so far as known, has never been tested to the limit; it is probable that the maximum is considerably in excess of what they have been called upon to show. In actual practice the average is a little over one pound for each one-half square foot of surface area.

Allowing that 600 square feet of surface will be used, the next question is how to distribute it to the best advantage. This is another important matter in which individual preference must rule. We have seen how the professionals disagree on this point, some using auxiliary planes of large size, and others depending upon smaller auxiliaries with an increase in number so as to secure on a different plan virtually the same amount of surface.

In deciding upon this feature the best thing to do is to follow the plans of some successful aviator, increasing the area of the auxiliaries in proportion to the increase in the area of the main planes. Thus, if you use 600 square feet of surface where the man whose plans you are following uses 500, it is simply a matter of making your planes one-fifth larger all around.

The Cost of Production.

Cost of production will be of interest to the amateur who essays to construct a flying machine. Assuming that the size decided upon is double that of the glider the material for the framework, timber, cloth, wire, etc., will cost a little more than double. This is because it must be heavier in proportion to the increased size of the framework, and heavy material brings a larger price than the lighter goods. If we allow $20 as the cost of the glider material it will be safe to put down the cost of that required for a real flying machine framework at $60, provided the owner builds it himself.

As regards the cost of motor and similar equipment it can only be said that this depends upon the selection made. There are some reliable aviation motors which may be had as low as $500, and there are others which cost as much as $2,000.

Services of Expert Necessary.

No matter what kind of a motor may be selected the services of an expert will be necessary in its proper installation unless the amateur has considerable genius in this line himself. As a general thing $25 should be a liberal allowance for this work. No matter how carefully the engine may be placed and connected it will be largely a matter of luck if it is installed in exactly the proper manner at the first attempt. The chances are that several alterations, prompted by the results of trials, will have to be made. If this is the case the expert's bill may readily run up to $50. If the amateur is competent to do this part of the work the entire item of $50 may, of course, be cut out.

As a general proposition a fairly satisfactory flying machine, one that will actually fly and carry the operator with it, may be constructed for $750, but it will lack the better qualities which mark the higher priced machines. This computation is made on the basis of $60 for material, $50 for services of expert, $600 for motor, etc., and an allowance of $40 for extras.

No man who has the flying machine germ in his system will be long satisfied with his first moderate price machine, no matter how well it may work. It's the old story of the automobile "bug" over again. The man who starts in with a modest $1,000 automobile invariably progresses by easy stages to the $4,000 or $5,000 class. The natural tendency is to want the biggest and best attainable within the financial reach of the owner.

It's exactly the same way with the flying machine convert. The more proficient he becomes in the manipulation of his car, the stronger becomes the desire to fly further and stay in the air longer than the rest of his brethren. This necessitates larger, more powerful, and more expensive machines as the work of the germ progresses.

Speed Affects Weight Capacity.

Don't overlook the fact that the greater speed you can attain the smaller will be the surface area you can get along with. If a machine with 500 square feet of sustaining surface, traveling at a speed of 40 miles an hour, will carry a weight of 1,200 pounds, we can cut the sustaining surface in half and get along with 250 square feet, provided a speed of 60 miles an hour can be obtained. At 100 miles an hour only 80 square feet of surface area would be required. In both instances the weight sustaining capacity will remain the same as with the 500 square feet of surface area—1,200 pounds.

One of these days some mathematical genius will figure out this problem with exactitude and we will have a dependable table giving the maximum carrying capacity of various surface areas at various stated speeds, based on the dimensions of the advancing edges. At present it is largely a matter of guesswork so far as making accurate computation goes. Much depends upon the shape of the machine, and the amount of surface offering resistance to the wind, etc.

CHAPTER IX. SELECTION OF THE MOTOR.

Motors for flying machines must be light in weight, of great strength, productive of extreme speed, and positively dependable in action. It matters little as to the particular form, or whether air or water cooled, so long as the four features named are secured. There are at least a dozen such motors or engines now in use. All are of the gasolene type, and all possess in greater or lesser degree the desired qualities. Some of these motors are:

Renault—8-cylinder, air-cooled; 50 horse power; weight 374 pounds.

Fiat—8-cylinder, air-cooled; 50 horse power; weight 150 pounds.

Farcot—8-cylinder, air-cooled; from 30 to 100 horse power, according to bore of cylinders; weight of smallest, 84 pounds.

R. E. P.—10-cylinder, air-cooled; 150 horse power; weight 215 pounds.

Gnome—7 and 14 cylinders, revolving type, air-cooled; 50 and 100 horse power; weight 150 and 300 pounds.

Darracq—2 to 14 cylinders, water cooled; 30 to 200 horse power; weight of smallest 100 pounds.

Wright—4-cylinder, water-cooled; 25 horse power; weight 200 pounds.

Antoinette—8 and 16-cylinder, water-cooled; 50 and 100 horse power; weight 250 and 500 pounds.

E. N. V.—8-cylinder, water-cooled; from 30 to 80 horse power, according to bore of cylinder; weight 150 to 400 pounds.

Curtiss—8-cylinder, water-cooled; 60 horse power; weight 300 pounds.

Average Weight Per Horse Power.

It will be noticed that the Gnome motor is unusually light, being about three pounds to the horse power produced, as opposed to an average of 4 1/2 pounds per horse power in other makes. This result is secured by the elimination of the fly-wheel, the engine itself revolving, thus obtaining the same effect that would be produced by a fly-wheel. The Farcot is even lighter, being considerably less than three pounds per horse power, which is the nearest approach to the long-sought engine equipment that will make possible a complete flying machine the total weight of which will not exceed one pound per square foot of area.

How Lightness Is Secured.

Thus far foreign manufacturers are ahead of Americans in the production of light-weight aerial motors, as is evidenced by the Gnome and Farcot engines, both of which are of French make. Extreme lightness is made possible by the use of fine, specially prepared steel for the cylinders, thus permitting them to be much thinner than if ordinary forms of steel were used. Another big saving in weight is made by substituting what are known as "auto lubricating" alloys for bearings. These alloys are made of a combination of aluminum and magnesium.

Still further gains are made in the use of alloy steel tubing instead of solid rods, and also by the paring away of material wherever it can be done without sacrificing strength. This plan, with the exclusive use of the best grades of steel, regardless of cost, makes possible a marked reduction in weight.

Multiplicity of Cylinders.

Strange as it may seem, multiplicity of cylinders does not always add proportionate weight. Because a 4-cylinder motor weighs say 100 pounds, it does not necessarily follow that an 8-cylinder equipment will weigh 200 pounds. The reason of this will be plain when it is understood that many of the parts essential to a 4-cylinder motor will fill the requirements of an 8-cylinder motor without enlargement or addition.

Neither does multiplying the cylinders always increase the horsepower proportionately. If a 4-cylinder motor is rated at 25 horsepower it is not safe to take it for granted that double the number of cylinders will give 50 horsepower. Generally speaking, eight cylinders, the bore, stroke and speed being the same, will give double the power that can be obtained from four, but this does not always hold good. Just why this exception should occur is not explainable by any accepted rule.

Horse Power and Speed.

Speed is an important requisite in a flying-machine motor, as the velocity of the aeroplane is a vital factor in flotation. At first thought, the propeller and similar adjuncts being equal, the inexperienced mind would naturally argue that a 50-horsepower engine should produce just double the speed of one of 25-horsepower. That this is a fallacy is shown by actual performances. The Wrights, using a 25-horsepower motor, have made 44 miles an hour, while Bleriot, with a 50-horsepower motor, has a record of a short-distance flight at the rate of 52 miles an hour. The fact is that, so far as speed is concerned, much depends upon the velocity of the wind, the size and shape of the aeroplane itself, and the size, shape and gearing of the propeller. The stronger the wind is blowing the easier it will be for the aeroplane to ascend, but at the same time the more difficult it will be to make headway against the wind in a horizontal direction. With a strong head wind, and proper engine force, your machine will progress to a certain extent, but it will be at an angle. If the aviator desired to keep on going upward this would be all right, but there is a limit to the altitude which it is desirable to reach—from 100 to 500 feet for experts—and after that it becomes a question of going straight ahead.

Great Waste of Power.

One thing is certain—even in the most efficient of modern aerial motors there is a great loss of power between the two points of production and effect. The Wright outfit, which is admittedly one of the most effective in use, takes one horsepower of force for the raising and propulsion of each 50 pounds of weight. This, for a 25-horsepower engine, would give a maximum lifting capacity of 1250 pounds. It is doubtful if any of the higher rated motors have greater efficiency. As an 8-cylinder motor requires more fuel to operate than a 4-cylinder, it naturally follows that it is more expensive to run than the smaller motor, and a normal increase in capacity, taking actual performances as a criterion, is lacking. In other words, what is the sense of using an 8-cylinder motor when one of 4 cylinders is sufficient?

What the Propeller Does.

Much of the efficiency of the motor is due to the form and gearing of the propeller. Here again, as in other vital parts of flying-machine mechanism, we have a wide divergence of opinion as to the best form. A fish makes progress through the water by using its fins and tail; a bird makes its way through the air in a similar manner by the use of its wings and tail. In both instances the motive power comes from the body of the fish or bird.

In place of fins or wings the flying machine is equipped with a propeller, the action of which is furnished by the engine. Fins and wings have been tried, but they don't work.

While operating on the same general principle, aerial propellers are much larger than those used on boats. This is because the boat propeller has a denser, more substantial medium to work in (water), and consequently can get a better "hold," and produce more propulsive force than one of the same size revolving in the air. This necessitates the aerial propellers being much larger than those employed for marine purposes. Up to this point all aviators agree, but as to the best form most of them differ.

Kinds of Propellers Used.

One of the most simple is that used by Curtiss. It consists of two pear-shaped blades of laminated wood, each blade being 5 inches wide at its extreme point, tapering slightly to the shaft connection. These blades are joined at the engine shaft, in a direct line. The propeller has a pitch of 5 feet, and weighs, complete, less than 10 pounds. The length from end to end of the two blades is 6 1/2 feet.

Wright uses two wooden propellers, in the rear of his biplane, revolving in opposite directions. Each propeller is two-bladed.

Bleriot also uses a two-blade wooden propeller, but it is placed in front of his machine. The blades are each about 3 1/2 feet long and have an acute "twist."

Santos-Dumont uses a two-blade wooden propeller, strikingly similar to the Bleriot.

On the Antoinette monoplane, with which good records have been made, the propeller consists of two spoon-shaped pieces of metal, joined at the engine shaft in front, and with the concave surfaces facing the machine.

The propeller on the Voisin biplane is also of metal, consisting of two aluminum blades connected by a forged steel arm.

Maximum thrust, or stress—exercise of the greatest air-displacing force—is the object sought. This, according to experts, is best obtained with a large propeller diameter and reasonably low speed. The diameter is the distance from end to end of the blades, which on the largest propellers ranges from 6 to 8 feet. The larger the blade surface the greater will be the volume of air displaced, and, following this, the greater will be the impulse which forces the aeroplane ahead. In all centrifugal motion there is more or less tendency to disintegration in the form of "flying off" from the center, and the larger the revolving object is the stronger is this tendency. This is illustrated in the many instances in which big grindstones and flywheels have burst from being revolved too fast. To have a propeller break apart in the air would jeopardize the life of the aviator, and to guard against this it has been found best to make its revolving action comparatively slow. Besides this the slow motion (it is only comparatively slow) gives the atmosphere a chance to refill the area disturbed by one propeller blade, and thus have a new surface for the next blade to act upon.

Placing of the Motor.

As on other points, aviators differ widely in their ideas as to the proper position for the motor. Wright locates his on the lower plane, midway between the front and rear edges, but considerably to one side of the exact center. He then counter-balances the engine weight by placing his seat far enough away in the opposite direction to preserve the center of gravity. This leaves a space in the center between the motor and the operator in which a passenger may be carried without disturbing the equilibrium.

Bleriot, on the contrary, has his motor directly in front and preserves the center of gravity by taking his seat well back, this, with the weight of the aeroplane, acting as a counter-balance.

On the Curtiss machine the motor is in the rear, the forward seat of the operator, and weight of the horizontal rudder and damping plane in front equalizing the engine weight.

No Perfect Motor as Yet.

Engine makers in the United States, England, France and Germany are all seeking to produce an ideal motor for aviation purposes. Many of the productions are highly creditable, but it may be truthfully said that none of them quite fill the bill as regards a combination of the minimum of weight with the maximum of reliable maintained power. They are all, in some respects, improvements upon those previously in use, but the great end sought for has not been fully attained.

One of the motors thus produced was made by the French firm of Darracq at the suggestion of Santos Dumont, and on lines laid down by him. Santos Dumont wanted a 2-cylinder horizontal motor capable of developing 30 horsepower, and not exceeding 4 1/2 pounds per horsepower in weight.

There can be no question as to the ability and skill of the Darracq people, or of their desire to produce a motor that would bring new credit and prominence to the firm. Neither could anything radically wrong be detected in the plans. But the motor, in at least one important requirement, fell short of expectations.

It could not be depended upon to deliver an energy of 30 horsepower continuously for any length of time. Its maximum power could be secured only in "spurts."

This tends to show how hard it is to produce an ideal motor for aviation purposes. Santos Dumont, of undoubted skill and experience as an aviator, outlined definitely what he wanted; one of the greatest designers in the business drew the plans, and the famous house of Darracq bent its best energies to the production. But the desired end was not fully attained.

Features of Darracq Motor.

Horizontal motors were practically abandoned some time ago in favor of the vertical type, but Santos Dumont had a logical reason for reverting to them. He wanted to secure a lower center of gravity than would be possible with a vertical engine. Theoretically his idea was correct as the horizontal motor lies flat, and therefore offers less resistance to the wind, but it did not work out as desired.

At the same time it must be admitted that this Darracq motor is a marvel of ingenuity and exquisite workmanship. The two cylinders, having a bore of 5 1-10 inches and a stroke of 4 7-10 inches, are machined out of a solid bar of steel until their weight is only 8 4-5 pounds complete. The head is separate, carrying the seatings for the inlet and exhaust valves, is screwed onto the cylinder, and then welded in position. A copper water-jacket is fitted, and it is in this condition that the weight of 8 4-5 pounds is obtained.

On long trips, especially in regions where gasolene is hard to get, the weight of the fuel supply is an important feature in aviation. As a natural consequence flying machine operators favor the motor of greatest economy in gasolene consumption, provided it gives the necessary power.

An American inventor, Ramsey by name, is working on a motor which is said to possess great possibilities in this line. Its distinctive features include a connecting rod much shorter than usual, and a crank shaft located the length of the crank from the central axis of the cylinder. This has the effect of increasing the piston stroke, and also of increasing the proportion of the crank circle during which effective pressure is applied to the crank.

Making the connecting rod shorter and leaving the crank mechanism the same would introduce excessive cylinder friction. This Ramsey overcomes by the location of his crank shaft. The effect of the long piston stroke thus secured, is to increase the expansion of the gases, which in turn increases the power of the engine without increasing the amount of fuel used.

Propeller Thrust Important.

There is one great principle in flying machine propulsion which must not be overlooked. No matter how powerful the engine may be unless the propeller thrust more than overcomes the wind pressure there can be no progress forward. Should the force of this propeller thrust and that of the wind pressure be equal the result is obvious. The machine is at a stand-still so far as forward progress is concerned and is deprived of the essential advancing movement.

Speed not only furnishes sustentation for the airship, but adds to the stability of the machine. An aeroplane which may be jerky and uncertain in its movements, so far as equilibrium is concerned, when moving at a slow gait, will readily maintain an even keel when the speed is increased.

Designs for Propeller Blades.

It is the object of all men who design propellers to obtain the maximum of thrust with the minimum expenditure of engine energy. With this purpose in view many peculiar forms of propeller blades have been evolved. In theory it would seem that the best effects could be secured with blades so shaped as to present a thin (or cutting) edge when they come out of the wind, and then at the climax of displacement afford a maximum of surface so as to displace as much air as possible. While this is the form most generally favored there are others in successful operation.

There is also wide difference in opinion as to the equipment of the propeller shaft with two or more blades. Some aviators use two and some four. All have more or less success. As a mathematical proposition it would seem that four blades should give more propulsive force than two, but here again comes in one of the puzzles of aviation, as this result is not always obtained.

Difference in Propeller Efficiency.

That there is a great difference in propeller efficiency is made readily apparent by the comparison of effects produced in two leading makes of machines—the Wright and the Voisin.

In the former a weight of from 1,100 to 1,200 pounds is sustained and advance progress made at the rate of 40 miles an hour and more, with half the engine speed of a 25 horse-power motor. This would be a sustaining capacity of 48 pounds per horsepower. But the actual capacity of the Wright machine, as already stated, is 50 pounds per horsepower.

The Voisin machine, with aviator, weighs about 1,370 pounds, and is operated with a so-horsepower motor. Allowing it the same speed as the Wright we find that, with double the engine energy, the lifting capacity is only 27 1/2 pounds per horsepower. To what shall we charge this remarkable difference? The surface of the planes is exactly the same in both machines so there is no advantage in the matter of supporting area.

Comparison of Two Designs.

On the Wright machine two wooden propellers of two blades each (each blade having a decided "twist") are used. As one 25 horsepower motor drives both propellers the engine energy amounts to just one-half of this for each, or 12 1/2 horsepower. And this energy is utilized at one-half the normal engine speed.

On the Voisin a radically different system is employed. Here we have one metal two-bladed propeller with a very slight "twist" to the blade surfaces. The full energy of a 50-horsepower motor is utilized.

Experts Fail to Agree.

Why should there be such a marked difference in the results obtained? Who knows? Some experts maintain that it is because there are two propellers on the Wright machine and only one on the Voisin, and consequently double the propulsive power is exerted. But this is not a fair deduction, unless both propellers are of the same size. Propulsive power depends upon the amount of air displaced, and the energy put into the thrust which displaces the air.

Other experts argue that the difference in results may be traced to the difference in blade design, especially in the matter of "twist."

The fact is that propeller results depend largely upon the nature of the aeroplanes on which they are used. A propeller, for instance, which gives excellent results on one type of aeroplane, will not work satisfactorily on another.

There are some features, however, which may be safely adopted in propeller selection. These are: As extensive a diameter as possible; blade area 10 to 15 per cent of the area swept; pitch four-fifths of the diameter; rotation slow. The maximum of thrust effort will be thus obtained.

CHAPTER X. PROPER DIMENSIONS OF MACHINES.

In laying out plans for a flying machine the first thing to decide upon is the size of the plane surfaces. The proportions of these must be based upon the load to be carried. This includes the total weight of the machine and equipment, and also the operator. This will be a rather difficult problem to figure out exactly, but practical approximate figures may be reached.

It is easy to get at the weight of the operator, motor and propeller, but the matter of determining, before they are constructed, what the planes, rudders, auxiliaries, etc., will weigh when completed is an intricate proposition. The best way is to take the dimensions of some successful machine and use them, making such alterations in a minor way as you may desire.

Dimensions of Leading Machines.

In the following tables will be found the details as to surface area, weight, power, etc., of the nine principal types of flying machines which are now prominently before the public:

MONOPLANES	Passengers	Surface area sq. feet	Spread in linear feet	Depth in linear feet	No. of Cylinders	Weight Without Operator	Horse Power	Propeller Diameter
Santos-Dumont	1	110	16.0	26.0	2	30	250	5.0
Bleriot	1	150.6	24.6	22.0	3	25	680	6.9
R. E. P.	1	215	34.1	28.9	7	35	900	6.6
Bleriot	2	236	32.9	23.0	7	50	1,240	8.1
Antoinette	2	538	41.2	37.9	8	50	1,040	7.2

BIPLANES	Passengers	Surface Area sq. feet	Spread in linear feet	Depth in linear feet	No. of Cylinders	Weight Without Operator	Horse Power	Propeller Diameter
Curtiss	2	258	29.0	28.7	8	50	600	6.0
Wright	2	538	41.0	30.7	4	25	1,100	8.1
Farman	2	430	32.9	39.6	7	50	1,200	8.9
Voisin	2	538	37.9	39.6	8	50	1,200	6.6

In giving the depth dimensions the length over all—from the extreme edge of the front auxiliary plane to the extreme tip of the rear is stated. Thus while the dimensions of the main planes of the

Wright machine are 41 feet spread by 6 1/2 feet in depth, the depth over all is 30.7.

Figuring Out the Details.

With this data as a guide it should be comparatively easy to decide upon the dimensions of the machine required. In arriving at the maximum lifting capacity the weight of the operator must be added. Assuming this to average 170 pounds the method of procedure would be as follows:

Add the weight of the operator to the weight of the complete machine. The new Wright machine complete weighs 900 pounds. This, plus 170, the weight of the operator, gives a total of 1,070 pounds. There are 538 square feet of supporting surface, or practically one square foot of surface area to each two pounds of load.

There are some machines, notably the Bleriot, in which the supporting power is much greater. In this latter instance we find a surface area of 150 1/2 square feet carrying a load of 680 plus 170, or an aggregate of 850 pounds. This is the equivalent of five pounds to the square foot. This ratio is phenomenally large, and should not be taken as a guide by amateurs.

The Matter of Passengers.

These deductions are based on each machine carrying one passenger, which is admittedly the limit at present of the monoplanes like those operated for record-making purposes by Santos-Dumont and Bleriot. The biplanes, however, have a two-passenger capacity, and this adds materially to the proportion of their weight-sustaining power as compared with the surface area. In the following statement all the machines are figured on the one-passenger basis. Curtiss and Wright have carried two passengers on numerous occasions, and an extra 170 pounds should therefore be added to the total weight carried, which would materially increase the capacity. Even with the two-passenger load the limit is by no means reached, but as experiments have gone no further it is impossible to make more accurate figures.

Average Proportions of Load.

It will be interesting, before proceeding to lay out the dimension details, to make a comparison of the proportion of load effect with the supporting surfaces of various well-known machines. Here are the figures:

Santos-Dumont—A trifle under four pounds per square foot.

Bleriot—Five pounds.

R. E. P.—Five pounds.

Antoinette—About two and one-quarter pounds.

Curtiss—About two and one-half pounds.

Wright—Two and one-quarter pounds.

Farman—A trifle over three pounds.

Voisin—A little under two and one-half pounds.

Importance of Engine Power.

While these figures are authentic, they are in a way misleading, as the important factor of engine power is not taken into consideration. Let us recall the fact that it is the engine power which keeps the machine in motion, and that it is only while in motion that the machine will remain suspended in the air. Hence, to attribute the support solely to the surface area is erroneous. True, that once under headway the planes contribute largely to the sustaining effect, and are absolutely essential in aerial navigation—the motor could not rise without them—still, when it comes to a question of weight-sustaining power, we must also figure on the engine capacity.

In the Wright machine, in which there is a lifting capacity of approximately 2 1/4 pounds to the square foot of surface area, an engine of only 25 horsepower is used. In the Curtiss, which has a lifting capacity of 2 1/2 pounds per square foot, the engine is of 50 horsepower. This is another of the peculiarities of aerial construction and navigation. Here we have a gain of 1/4 pound in weight-lifting capacity with an expenditure of double the horsepower. It is this feature which enables Curtiss to get along with a smaller surface area of supporting planes at the expense of a big increase in engine power. Proper Weight of Machine.

As a general proposition the most satisfactory machine for amateur purposes will be found to be one with a total weight-sustaining power of about 1,200 pounds. Deducting 170 pounds as the weight of the operator, this will leave 1,030 pounds for the complete motor-equipped machine, and it should be easy to construct one within this limit. This implies, of course, that due care will be taken to eliminate all superfluous weight by using the lightest material compatible with strength and safety.

This plan will admit of 686 pounds weight in the frame work, coverings, etc., and 344 for the motor, propeller, etc., which will be ample. Just how to distribute the weight of the planes is a matter which must be left to the ingenuity of the builder.

Comparison of Bird Power.

There is an interesting study in the accompanying illustration. Note that the surface area of the albatross is much smaller than that of the vulture, although the wing spread is about the same. Despite this the albatross accomplishes fully as much in the way of flight and soaring as the vulture. Why? Because the albaboss is quicker and more powerful in action. It is the application of this same principle in flying machines which enables those of great speed and power to get along with less supporting surface than those of slower movement.

Measurements of Curtiss Machine.

Some idea of framework proportion may be had from the following description of the Curtiss machine. The main planes have a spread (width) of 29 feet, and are 4 1/2 feet deep. The front double surface horizontal rudder is 6x2 feet, with an area of 24 square feet. To the rear of the main planes is a single surface horizontal plane 6x2 feet, with an area of 12 square feet. In connection with this is a vertical rudder 2 1/2 feet square. Two movable ailerons, or balancing planes, are placed at the extreme ends of the upper planes. These are 6x2 feet, and have a combined area of 24 square feet. There is also a triangular shaped vertical steadying surface in connection with the front rudder.

Thus we have a total of 195 square feet, but as the official figures are 258, and the size of the triangular-shaped steadying surface is unknown, we must take it for granted that this makes up the difference. In the matter of proportion the horizontal double-plane rudder is about one-tenth the size of the main plane, counting the surface area of only one plane, the vertical rudder one-fortieth, and the ailerons one-twentieth.

CHAPTER XI. PLANE AND RUDDER CONTROL.

Having constructed and equipped your machine, the next thing is to decide upon the method of controlling the various rudders and auxiliary planes by which the direction and equilibrium and ascending and descending of the machine are governed.

The operator must be in position to shift instantaneously the position of rudders and planes, and also to control the action of the motor. This latter is supposed to work automatically and as a general thing does so with entire satisfaction, but there are times when the supply of gasolene must be regulated, and similar things done. Airship navigation calls for quick action, and for this reason the matter of control is an important one—it is more than important; it is vital.

Several Methods of Control.

Some aviators use a steering wheel somewhat after the style of that used in automobiles, and by this not only manipulate the rudder planes, but also the flow of gasolene. Others employ foot levers, and still others, like the Wrights, depend upon hand levers.

Curtiss steers his aeroplane by means of a wheel, but secures the desired stabilizing effect with an ingenious jointed chair-back. This is so arranged that by leaning toward the high point of his wing planes the aeroplane is restored to an even keel. The steering post of the wheel is movable backward and forward, and by this motion elevation is obtained.

The Wrights for some time used two hand levers, one to steer by and warp the flexible tips of the planes, the other to secure elevation. They have now consolidated all the functions in one lever. Bleriot also uses the single lever control.

Farman employs a lever to actuate the rudders, but manipulates the balancing planes by foot levers.

Santos-Dumont uses two hand levers with which to steer and elevate, but manipulates the planes by means of an attachment to the back of his outer coat.

Connection With the Levers.

No matter which particular method is employed, the connection between the levers and the object to be manipulated is almost invariably by wire. For instance, from the steering levers (or lever) two wires connect with opposite sides of the rudder. As a lever is moved so as to draw in the right-hand wire the rudder is drawn to the right and vice versa. The operation is exactly the same as in steering a boat. It is the same way in changing the position of the balancing planes. A movement of the hands or feet and the machine has changed its course, or, if the equilibrium is threatened, is back on an even keel.

Simple as this seems it calls for a cool head, quick eye, and steady hand. The least hesitation or a false movement, and both aviator and craft are in danger.

Which Method is Best?

It would be a bold man who would attempt to pick out any one of these methods of control and say it was better than the others. As in other sections of aeroplane mechanism each method has its advocates who dwell learnedly upon its advantages, but the fact remains that all the various plans work well and give satisfaction.

What the novice is interested in knowing is how the control is effected, and whether he has become proficient enough in his manipulation of it to be absolutely dependable in time of emergency. No amateur should attempt a flight alone, until he has thoroughly mastered the steering and plane control. If the services and advice of an experienced aviator are not to be had the novice should mount his machine on some suitable supports so it will be well clear of the ground, and, getting into the operator's seat, proceed to make himself well acquainted with the operation of the steering wheel and levers.

Some Things to Be Learned.

He will soon learn that certain movements of the steering gear produce certain effects on the rudders. If, for instance, his machine is equipped with a steering wheel, he will find that turning the wheel to the right turns the aeroplane in the same direction, because the tiller is brought around to the left. In the same way he will learn that a given movement of the lever throws the forward edge of the main plane upward, and that the machine, getting the impetus of the wind under the concave surfaces of the planes, will ascend. In the same way it will quickly become apparent to him that an opposite movement of the lever will produce an opposite effect—the forward edges of the planes will be lowered, the air will be "spilled" out to the rear, and the machine will descend.

The time expended in these preliminary lessons will be well spent. It would be an act of folly to attempt to actually sail the craft without them.

CHAPTER XII. HOW TO USE THE MACHINE.

It is a mistaken idea that flying machines must be operated at extreme altitudes. True, under the impetus of handsome prizes, and the incentive to advance scientific knowledge, professional aviators have ascended to considerable heights, flights at from 500 to 1,500 feet being now common with such experts as Farman, Bleriot, Latham, Paulhan, Wright and Curtiss. The altitude record at this time is about 4,165 feet, held by Paulhan.

One of the instructions given by experienced aviators to pupils, and for which they insist upon implicit obeyance, is: "If your machine gets more than 30 feet high, or comes closer to the ground than 6 feet, descend at once." Such men as Wright and Curtiss will not tolerate a violation of this rule. If their instructions are not strictly complied with they decline to give the offender further lessons.

Why This Rule Prevails.

There is good reason for this precaution. The higher the altitude the more rarefied (thinner) becomes the air, and the less sustaining power it has. Consequently the more difficult it becomes to keep in suspension a given weight. When sailing within 30 feet of the ground sustentation is comparatively easy and, should a fall occur, the results are not likely to be serious. On the other hand, sailing too near the ground is almost as objectionable in many ways as getting up too high. If the craft is navigated too close to the ground trees, shrubs, fences and other obstructions are liable to be encountered. There is also the handicap of contrary air currents diverted by the obstructions referred to, and which will be explained more fully further on.

How to Make a Start.

Taking it for granted that the beginner has familiarized himself with the manipulation of the machine, and especially the control mechanism, the next thing in order is an actual flight. It is probable that his machine will be equipped with a wheeled alighting gear, as the skids used by the Wrights necessitate the use of a special starting track. In this respect the wheeled machine is much easier to handle so far as novices are concerned as it may be easily rolled to the trial grounds. This, as in the case of the initial experiments, should be a clear, reasonably level place, free from trees, fences, rocks and similar obstructions with which there may be danger of colliding.

The beginner will need the assistance of three men. One of these should take his position in the rear of the machine, and one at each end. On reaching the trial ground the aviator takes his seat in the machine and, while the men at the ends hold it steady the one in the rear assists in retaining it until the operator is ready. In the meantime the aviator has started his motor. Like the glider the flying machine, in order to accomplish the desired results, should be headed into the wind.

When the Machine Rises.

Under the impulse of the pushing movement, and assisted by the motor action, the machine will gradually rise from the ground—provided it has been properly proportioned and put together, and everything is in working order. This is the time when the aviator requires a cool head, At a modest distance from the ground use the control lever to bring the machine on a horizontal level and overcome the tendency to rise. The exact manipulation of this lever depends upon the method of control adopted, and with this the aviator is supposed to have thoroughly familiarized himself as previously advised in Chapter XI.

It is at this juncture that the operator must act promptly, but with the perfect composure begotten of confidence. One of the great drawbacks in aviation by novices is the tendency to become rattled, and this is much more prevalent than one might suppose, even among men who, under other conditions, are cool and confident in their actions.

There is something in the sensation of being suddenly lifted from the ground, and suspended in the air that is disconcerting at the start, but this will soon wear off if the experimenter will keep cool. A few successful flights no matter how short they may be, will put a lot of confidence into him.

Make Your Flights Short.

Be modest in your initial flights. Don't attempt to match the records of experienced men who have devoted years to mastering the details of aviation. Paulhan, Farman, Bleriot, Wright, Curtiss, and all the rest of them began, and practiced for years, in the manner here described, being content to make just a little advancement at each attempt. A flight of 150 feet, cleanly and safely made, is better as a beginning than one of 400 yards full of bungling mishaps.

And yet these latter have their uses, provided the operator is of a discerning mind and can take advantage of them as object lessons. But, it is not well to invite them. They will occur frequently enough under the most favorable conditions, and it is best to have them come later when the feeling of trepidation and uncertainty as to what to do has worn off.

Above all, don't attempt to fly too high. Keep within a reasonable distance from the ground—about 25 or 30 feet. This advice is not given solely to lessen the risk of serious accident in case of collapse, but mainly because it will assist to instill confidence in the operator.

It is comparatively easy to learn to swim in shallow water, but the knowledge that one is tempting death in deep water begets timidity.

Preserving the Equilibrium.

After learning how to start and stop, to ascend and descend, the next thing to master is the art of preserving equilibrium, the knack of keeping the machine perfectly level in the air—on an "even keel," as a sailor would say. This simile is particularly appropriate as all aviators are in reality sailors, and much more daring ones than those who course the seas. The latter are in craft which are kept afloat by the buoyancy of the water, whether in motion or otherwise and, so long as normal conditions prevail, will not sink. Aviators sail the air in craft in which constant motion must be maintained in order to ensure flotation.

The man who has ridden a bicycle or motorcycle around curves at anything like high speed, will have a very good idea as to the principle of maintaining equilibrium in an airship. He knows that in rounding curves rapidly there is a marked tendency to change the direction of the motion which will result in an upset unless he overcomes it by an inclination of his body in an opposite direction. This is why we see racers lean well over when taking the curves. It simply must be done to preserve the equilibrium and avoid a spill.

How It Works In the Air.

If the equilibrium of an airship is disturbed to an extent which completely overcomes the center of gravity it falls according to the location of the displacement. If this displacement, for instance, is at either end the apparatus falls endways; if it is to the front or rear, the fall is in the corresponding direction.

Owing to uncertain air currents—the air is continually shifting and eddying, especially within a hundred feet or so of the earth—the equilibrium of an airship is almost constantly being disturbed to some extent. Even if this disturbance is not serious enough to bring on a fall it interferes with the progress of the machine, and should be overcome at once. This is one of the things connected with aerial navigation which calls for prompt, intelligent action.

Frequently, when the displacement is very slight, it may be overcome, and the craft immediately righted by a mere shifting of the operator's body. Take, for illustration, a case in which the extreme right end of the machine becomes lowered a trifle from the normal level. It is possible to bring it back into proper position by leaning over to the left far enough to shift the weight to the counterbalancing point. The same holds good as to minor front or rear displacements.

When Planes Must Be Used.

There are other displacements, however, and these are the most frequent, which can be only overcome by manipulation of the stabilizing planes. The method of procedure depends upon the form of machine in use. The Wright machine, as previously explained, is equipped with plane ends which are so contrived as to admit of their being warped (position changed) by means of the lever control. These flexible tip planes move simultaneously, but in opposite directions. As those on one end rise, those on the other end fall below the level of the main plane. By this means air is displaced at one point, and an increased amount secured in another.

This may seem like a complicated system, but its workings are simple when once understood. It is by the manipulation or warping of these flexible tips that transverse stability is maintained, and any tendency to displacement endways is overcome. Longitudinal stability is governed by means of the front rudder.

Stabilizing planes of some form are a feature, and a necessary feature, on all flying machines, but the methods of application and manipulation vary according to the individual ideas of the inventors. They all tend, however, toward the same end—the keeping of the machine perfectly level when being navigated in the air.

When to Make a Flight.

A beginner should never attempt to make a flight when a strong wind is blowing. The fiercer the wind, the more likely it is to be gusty and uncertain, and the more difficult it will be to control the machine. Even the most experienced and daring of aviators find there is a limit to wind speed against which they dare not compete. This is not because they lack courage, but have the sense to realize that it would be silly and useless.

The novice will find a comparatively still day, or one when the wind is blowing at not to exceed 15 miles an hour, the best for his experiments. The machine will be more easily controlled, the trip will be safer, and also cheaper as the consumption of fuel increases with the speed of the wind against which the aeroplane is forced.

CHAPTER XIII. PECULIARITIES OF AIRSHIP POWER.

As a general proposition it takes much more power to propel an airship a given number of miles in a certain time than it does an automobile carrying a far heavier load. Automobiles with a gross load of 4,000 pounds, and equipped with engines of 30 horsepower, have travelled considerable distances at the rate of 50 miles an hour. This is an equivalent of about 134 pounds per horsepower. For an average modern flying machine, with a total load, machine and passengers, of 1,200 pounds, and equipped with a 50-horsepower engine, 50 miles an hour is the maximum. Here we have the equivalent of exactly 24 pounds per horsepower. Why this great difference?

No less an authority than Mr. Octave Chanute answers the question in a plain, easily understood manner. He says:

"In the case of an automobile the ground furnishes a stable support; in the case of a flying machine the engine must furnish the support and also velocity by which the apparatus is sustained in the air."

Pressure of the Wind.

Air pressure is a big factor in the matter of aeroplane horsepower. Allowing that a dead calm exists, a body moving in the atmosphere creates more or less resistance. The faster it moves, the greater is this resistance. Moving at the rate of 60 miles an hour the resistance, or wind pressure, is approximately 50 pounds to the square foot of surface presented. If the moving object is advancing at a right angle to the wind the following table will give the horsepower effect of the resistance per square foot of surface at various speeds.

Horse Power Miles per Hour per sq. foot 10 0.013 15 0 044 20 0.105 25 0.205 30 0.354 40 0.84 50 1.64 60 2.83 80 6.72 100 13.12

While the pressure per square foot at 60 miles an hour, is only 1.64 horsepower, at 100 miles, less than double the speed, it has increased to 13.12 horsepower, or exactly eight times as much. In other words the pressure of the wind increases with the square of the velocity. Wind at 10 miles an hour has four times more pressure than wind at 5 miles an hour.

How to Determine Upon Power.

This element of air resistance must be taken into consideration in determining the engine horsepower required. When the machine is under headway sufficient to raise it from the ground (about 20 miles an hour), each square foot of surface resistance, will require nearly nine-tenths of a horsepower to overcome the wind pressure, and propel the machine through the air. As shown in the table the ratio of power required increases rapidly as the speed increases until at 60 miles an hour approximately 3 horsepower is needed.

In a machine like the Curtiss the area of wind-exposed surface is about 15 square feet. On the basis of this resistance moving the machine at 40 miles an hour would require 12 horsepower. This computation covers only the machine's power to overcome resistance. It does not cover the power exerted in propelling the machine forward after the air pressure is overcome. To meet this important requirement Mr. Curtiss finds it necessary to use a 50-horsepower engine. Of this power, as has been already stated, 12 horsepower is consumed in meeting the wind pressure, leaving 38 horsepower for the purpose of making progress.

The flying machine must move faster than the air to which it is opposed. Unless it does this there can be no direct progress. If the two forces are equal there is no straight-ahead advancement. Take, for sake of illustration, a case in which an aeroplane, which has developed a speed of 30 miles an hour, meets a wind velocity of equal force moving in an opposite direction. What is the result? There can be no advance because it is a contest between two evenly matched forces. The aeroplane stands still. The only way to get out of the difficulty is for the operator to wait for more favorable conditions, or bring his machine to the ground in the usual manner by manipulation of the control system.

Take another case. An aeroplane, capable of making 50 miles an hour in a calm, is met by a head wind of 25 miles an hour. How much progress does the aeroplane make? Obviously it is 25 miles an hour over the ground.

Put the proposition in still another way. If the wind is blowing harder than it is possible for the engine power to overcome, the machine will be forced backward.

Wind Pressure a Necessity.

While all this is true, the fact remains that wind pressure, up to a certain stage, is an absolute necessity in aerial navigation. The atmosphere itself has very little real supporting power, especially if inactive. If a body heavier than air is to remain afloat it must move rapidly while in suspension.

One of the best illustrations of this is to be found in skating over thin ice. Every school boy knows that if he moves with speed he may skate or glide in safety across a thin sheet of ice that would not begin to bear his weight if he were standing still. Exactly the same proposition obtains in the case of the flying machine.

The non-technical reason why the support of the machine becomes easier as the speed increases is that the sustaining power of the atmosphere increases with the resistance, and the speed with which the object is moving increases this resistance. With a velocity of 12 miles an hour the weight of the machine is practically reduced by 230 pounds. Thus, if under a condition of absolute calm it were possible to sustain a weight of 770 pounds, the same atmosphere would sustain a weight of 1,000 pounds moving at a speed of 12 miles an hour. This sustaining power increases rapidly as the speed increases. While at 12 miles the sustaining power is figured at 230 pounds, at 24 miles it is four times as great, or 920 pounds.

Supporting Area of Birds.

One of the things which all producing aviators seek to copy is the motive power of birds, particularly in their relation to the area of support. Close investigation has established the fact that the larger the bird the less is the relative area of support required to secure a given result. This is shown in the following table:

Supporting Weight Surface Horse area Bird in lbs. in sq. feet power per lb. Pigeon 1.00 0.7 0.012 0.7 Wild Goose 9.00 2.65 0.026 0.2833 Buzzard 5.00 5.03 0.015 1.06 Condor 17.00 9.85 0.043 0.57

So far as known the condor is the largest of modern birds. It has a wing stretch of 10 feet from tip to tip, a supporting area of about 10 square feet, and weighs 17 pounds. It. is capable of exerting perhaps 1-30 horsepower. (These figures are, of course, approximate.) Comparing the condor with the buzzard with a wing stretch of 6 feet, supporting area of 5 square feet, and a little over 1-100 horsepower, it may be seen that, broadly speaking, the larger the bird the less surface area (relatively) is needed for its support in the air.

Comparison With Aeroplanes.

If we compare the bird figures with those made possible by the development of the aeroplane it will be readily seen that man has made a wonderful advance in imitating the results produced by nature. Here are the figures:

Machine	Supporting Weight in lbs.	Surface area in sq. feet	Horse power	power per lb.
Santos-Dumont	350	110.00	30	0.314
Bleriot	700	150.00	25	0.214
Antoinette	1,200	538.00	50	0.448
Curtiss	700	258.00	60	0.368
Wright	4 1,100	538.00	25	0.489
Farman	1,200	430.00	50	0.358
Voisin	1,200	538.00	50	0.448

While the average supporting surface is in favor of the aeroplane, this is more than overbalanced by the greater amount of horsepower required for the weight lifted. The average supporting surface in birds is about three-quarters of a square foot per pound. In the average aeroplane it is about one-half square foot per pound. On the other hand the average aeroplane has a lifting capacity of 24 pounds per horsepower, while the buzzard, for instance, lifts 5 pounds with 15-100 of a horsepower. If the Wright machine—which has a lifting power of 50 pounds per horsepower—should be alone considered the showing would be much more favorable to the aeroplane, but it would not be a fair comparison.

More Surface, Less Power.

Broadly speaking, the larger the supporting area the less will be the power required. Wright, by the use of 538 square feet of supporting surface, gets along with an engine of 25 horsepower. Curtiss, who uses only 258 square feet of surface, finds an engine of

50 horsepower is needed. Other things, such as frame, etc., being equal, it stands to reason that a reduction in the area of supporting surface will correspondingly reduce the weight of the machine. Thus we have the Curtiss machine with its 258 square feet of surface, weighing only 600 pounds (without operator), but requiring double the horsepower of the Wright machine with 538 square feet of surface and weighing 1,100 pounds. This demonstrates in a forceful way the proposition that the larger the surface the less power will be needed.

But there is a limit, on account of its bulk and awkwardness in handling, beyond which the surface area cannot be enlarged. Otherwise it might be possible to equip and operate aeroplanes satisfactorily with engines of 15 horsepower, or even less.

The Fuel Consumption Problem.

Fuel consumption is a prime factor in the production of engine power. The veriest mechanical tyro knows in a general way that the more power is secured the more fuel must be consumed, allowing that there is no difference in the power-producing qualities of the material used. But few of us understand just what the ratio of increase is, or how it is caused. This proposition is one of keen interest in connection with aviation.

Let us cite a problem which will illustrate the point quoted: Allowing that it takes a given amount of gasolene to propel a flying machine a given distance, half the way with the wind, and half against it, the wind blowing at one-half the speed of the machine, what will be the increase in fuel consumption?

Increase of Thirty Per Cent.

On the face of it there would seem to be no call for an increase as the resistance met when going against the wind is apparently offset by the propulsive force of the wind when the machine is travelling with it. This, however, is called faulty reasoning. The increase in fuel consumption, as figured by Mr. F. W. Lanchester, of the Royal Society of Arts, will be fully 30 per cent over the amount required for a similar operation of the machine in still air. If the journey should be made at right angles to the wind under the same conditions the increase would be 15 per cent.

In other words Mr. Lanchester maintains that the work done by the motor in making headway against the wind for a certain distance calls for more engine energy, and consequently more fuel by 30 per cent, than is saved by the helping force of the wind on the return journey.

CHAPTER XIV. ABOUT WIND CURRENTS, ETC.

One of the first difficulties which the novice will encounter is the uncertainty of the wind currents. With a low velocity the wind, some distance away from the ground, is ordinarily steady. As the velocity increases, however, the wind generally becomes gusty and fitful in its action. This, it should be remembered, does not refer to the velocity of the machine, but to that of the air itself.

In this connection Mr. Arthur T. Atherholt, president of the Aero Club of Pennsylvania, in addressing the Boston Society of Scientific Research, said:

"Probably the whirlpools of Niagara contain no more erratic currents than the strata of air which is now immediately above us, a fact hard to realize on account of its invisibility."

Changes In Wind Currents.

While Mr. Atherholt's experience has been mainly with balloons it is all the more valuable on this account, as the balloons were at the mercy of the wind and their varying directions afforded an indisputable guide as to the changing course of the air currents. In speaking of this he said:

"In the many trips taken, varying in distance traversed from twenty-five to 900 miles, it was never possible except in one instance to maintain a straight course. These uncertain currents were most noticeable in the Gordon-Bennett race from St. Louis in 1907. Of the nine aerostats competing in that event, eight covered a more or less direct course due east and southeast, whereas the writer, with Major Henry B. Hersey, first started northwest, then north, northeast, east, east by south, and when over the center of Lake Erie were again blown northwest notwithstanding that more favorable winds were sought for at altitudes varying from 100 to 3,000 meters, necessitating a finish in Canada nearly northeast of the starting point.

"These nine balloons, making landings extending from Lake Ontario, Canada, to Virginia, all started from one point within the same hour.

"The single exception to these roving currents occurred on October 21st, of last year (1909) when, starting from Philadelphia, the wind shifted more than eight degrees, the greatest variation being at the lowest altitudes, yet at no time was a height of over a mile reached.

"Throughout the entire day the sky was overcast, with a thermometer varying from fifty-seven degrees at 300 feet to forty-four degrees, Fahrenheit at 5,000 feet, at which altitude the wind had a velocity of 43 miles an hour, in clouds of a cirro-cumulus nature, a landing finally being made near Tannersville, New York, in the Catskill mountains, after a voyage of five and one-half hours.

"I have no knowledge of a recorded trip of this distance and duration, maintained in practically a straight line from start to finish."

This wind disturbance is more noticeable and more difficult to contend with in a balloon than in a flying machine, owing to the bulk and unwieldy character of the former. At the same time it is not conducive to pleasant, safe or satisfactory sky-sailing in an aeroplane. This is not stated with the purpose of discouraging aviation, but merely that the operator may know what to expect and be prepared to meet it.

Not only does the wind change its horizontal course abruptly and without notice, but it also shifts in a vertical direction, one second blowing up, and another down. No man has as yet fathomed the why and wherefore of this erratic action; it is only known that it exists.

The most stable currents will be found from 50 to 100 feet from the earth, provided the wind is not diverted by such objects as trees, rocks, etc. That there are equally stable currents higher up is true, but they are generally to be found at excessive altitudes.

How a Bird Meets Currents.

Observe a bird in action on a windy day and you will find it continually changing the position of its wings. This is done to meet the varying gusts and eddies of the air so that sustentation may be maintained and headway made. One second the bird is bending its wings, altering the angle of incidence; the next it is lifting or depressing one wing at a time. Still again it will extend one wing tip in advance of the other, or be spreading or folding, lowering or raising its tail.

All these motions have a meaning, a purpose. They assist the bird in preserving its equilibrium. Without them the bird would be just as helpless in the air as a human being and could not remain afloat.

When the wind is still, or comparatively so, a bird, having secured the desired altitude by flight at an angle, may sail or soar with no wing action beyond an occasional stroke when it desires to advance. But, in a gusty, uncertain wind it must use its wings or alight somewhere.

Trying to Imitate the Bird.

Writing in *Fly*, Mr. William E. White says:

"The bird's flight suggests a number of ways in which the equilibrium of a mechanical bird may be controlled. Each of these methods of control may be effected by several different forms of mechanism.

"Placing the two wings of an aeroplane at an angle of three to five degrees to each other is perhaps the oldest way of securing lateral balance. This way readily occurs to anyone who watches a sea gull soaring. The theory of the dihedral angle is that when one wing is lifted by a gust of wind, the air is spilled from under it; while the other wing, being correspondingly depressed, presents a greater resistance to the gust and is lifted restoring the balance. A fixed angle of three to five degrees, however, will only be sufficient for very light puffs of wind and to mount the wings so that the whole wing may be moved to change the dihedral angle presents mechanical difficulties which would be better avoided.

"The objection of mechanical impracticability applies to any plan to preserve the balance by shifting weight or ballast. The center of gravity should be lower than the center of the supporting surfaces, but cannot be made much lower. It is a common mistake to assume that complete stability will be secured by hanging the center of gravity very low on the principle of the parachute. An aeroplane depends upon rapid horizontal motion for its support, and if the center of gravity be far below the center of support, every change of speed or wind pressure will cause the machine to turn about its center of gravity, pitching forward and backward dangerously.

Preserving Longitudinal Balance.

"The birds maintain longitudinal, or fore and aft balance, by elevating or depressing their tails. Whether this action is secured in an aeroplane by means of a horizontal rudder placed in the rear, or by deflecting planes placed in front of the main planes, the principle is evidently the same. A horizontal rudder placed well to the rear as in the Antoinette, Bleriot or Santos-Dumont monoplanes, will be very much safer and steadier than the deflecting planes in front, as in the Wright or Curtiss biplanes, but not so sensitive or prompt in action.

"The natural fore and aft stability is very much strengthened by placing the load well forward. The center of gravity near the front and a tail or rudder streaming to the rear secures stability as an arrow is balanced by the head and feathering. The adoption of this principle makes it almost impossible for the aeroplane to turn over.

The Matter of Lateral Balance.

"All successful aeroplanes thus far have maintained lateral balance by the principle of changing the angle of incidence of the wings.

"Other ways of maintaining the lateral balance, suggested by observation of the flight of birds are—extending the wing tips and spilling the air through the pinions; or, what is the same thing, varying the area of the wings at their extremities.

"Extending the wing tips seems to be a simple and effective solution of the problem. The tips may be made to swing outward upon a vertical axis placed at the front edge of the main planes; or they may be hinged to the ends of the main plane so as to be elevated or depressed through suitable connections by the aviator; or they may be supported from a horizontal axis parallel with the ends of the main planes so that they may swing outward, the aviator controlling both tips through one lever so that as one tip is extended the other is retracted.

"The elastic wing pinions of a bird bend easily before the wind, permitting the gusts to glance off, but presenting always an even and efficient curvature to the steady currents of the air."

High Winds Threaten Stability.

To ensure perfect stability, without control, either human or automatic, it is asserted that the aeroplane must move faster than the wind is blowing. So long as the wind is blowing at the rate of 30 miles an hour, and the machine is traveling 40 or more, there will be little trouble as regards equilibrium so far as wind disturbance goes, provided the wind blows evenly and does not come in gusts or eddying currents. But when conditions are reversed—when the machine travels only 30 miles an hour and the wind blows at the rate of 50, look out for loss of equilibrium.

One of the main reasons for this is that high winds are rarely steady; they seldom blow for any length of time at the same speed. They are usually "gusty," the gusts being a momentary movement at a higher speed. Tornadic gusts are also formed by the meeting of two opposing currents, causing a whirling motion, which makes stability uncertain. Besides, it is not unusual for wind of high speed to suddenly change its direction without warning.

Trouble With Vertical Columns.

Vertical currents—columns of ascending air—are frequently encountered in unexpected places and have more or less tendency, according to their strength, to make it difficult to keep the machine within a reasonable distance from the ground.

These vertical currents are most generally noticeable in the vicinity of steep cliffs, or deep ravines. In such instances they are usually of considerable strength, being caused by the deflection of strong winds blowing against the face of the cliffs. This deflection exerts a back pressure which is felt quite a distance away from the point of origin, so that the vertical current exerts an influence in forcing the machine upward long before the cliff is reached.

CHAPTER XV. THE ELEMENT OF DANGER.

That there is an element of danger in aviation is undeniable, but it is nowhere so great as the public imagines. Men are killed and injured in the operation of flying machines just as they are killed and injured in the operation of railways. Considering the character of aviation the percentage of casualties is surprisingly small.

This is because the results following a collapse in the air are very much different from what might be imagined. Instead of dropping to the ground like a bullet an aeroplane, under ordinary conditions will, when anything goes wrong, sail gently downward like a parachute, particularly if the operator is cool-headed and nervy enough to so manipulate the apparatus as to preserve its equilibrium and keep the machine on an even keel.

Two Fields of Safety.

At least one prominent aviator has declared that there are two fields of safety—one close to the ground, and the other well up in the air. In the first-named the fall will be a slight one with little chance of the operator being seriously hurt. From the field of high altitude the the descent will be gradual, as a rule, the planes of the machine serving to break the force of the fall. With a cool-headed operator in control the aeroplane may be even guided at an angle (about 1 to 8) in its descent so as to touch the ground with a gliding motion and with a minimum of impact.

Such an experience, of course, is far from pleasant, but it is by no means so dangerous as might appear. There is more real danger in falling from an elevation of 75 or 100 feet than there is from 1,000 feet, as in the former case there is no chance for the machine to serve as a parachute—its contact with the ground comes too quickly.

Lesson in Recent Accidents.

Among the more recent fatalities in aviation are the deaths of Antonio Fernandez and Leon Delagrange. The former was thrown to the ground by a sudden stoppage of his motor, the entire machine seeming to collapse. It is evident there were radical defects, not only in the motor, but in the aeroplane framework as well. At the time of the stoppage it is estimated that Fernandez was up about 1,500 feet, but the machine got no opportunity to exert a parachute effect, as it broke up immediately. This would indicate a fatal weakness in the structure which, under proper testing, could probably have been detected before it was used in flight.

It is hard to say it, but Delagrange appears to have been culpable to great degree in overloading his machine with a motor equipment much heavier than it was designed to sustain. He was 65 feet up in the air when the collapse occurred, resulting in his death. As in the case of Fernandez common-sense precaution would doubtless have prevented the fatality.

Aviation Not Extra Hazardous.

All told there have been, up to the time of this writing (April, 1910), just five fatalities in the history of power-driven aviation. This is surprisingly low when the nature of the experiments, and the fact that most of the operators were far from having extended experience, is taken into consideration. Men like the Wrights, Curtiss, Bleriot, Farman, Paulhan and others, are now experts, but there was a time, and it was not long ago, when they were unskilled. That they, with numerous others less widely known, should have come safely through their many experiments would seem to disprove the prevailing idea that aviation is an extra hazardous pursuit.

In the hands of careful, quick-witted, nervy men the sailing of an airship should be no more hazardous than the sailing of a yacht. A vessel captain with common sense will not go to sea in a storm, or navigate a weak, unseaworthy craft. Neither should an aviator attempt to sail when the wind is high and gusty, nor with a machine which has not been thoroughly tested and found to be strong and safe.

Safer Than Railroading.

Statistics show that some 12,000 people are killed and 72,000 injured every year on the railroads of the United States. Come to think it over it is small wonder that the list of fatalities is so large. Trains are run at high speeds, dashing over crossings at which collisions are liable to occur, and over bridges which often collapse or are swept away by floods. Still, while the number of casualties is large, the actual percentage is small considering the immense number of people involved.

It is so in aviation. The number of casualties is remarkably small in comparison with the number of flights made. In the hands of competent men the sailing of an airship should be, and is, freer from risk of accident than the running of a railway train. There are no rails to spread or break, no bridges to collapse, no crossings at which collisions may occur, no chance for some sleepy or overworked employee to misunderstand the dispatcher's orders and cause a wreck.

Two Main Causes of Trouble.

The two main causes of trouble in an airship leading to disaster may be attributed to the stoppage of the motor, and the aviator becoming rattled so that he loses control of his machine. Modern ingenuity is fast developing motors that almost daily become more and more reliable, and experience is making aviators more and more self-confident in their ability to act wisely and promptly in cases of emergency. Besides this a satisfactory system of automatic control is in a fair way of being perfected.

Occasionally even the most experienced and competent of men in all callings become careless and by foolish action invite disaster. This is true of aviators the same as it is of railroaders, men who work in dynamite mills, etc. But in nearly every instance the responsibility rests with the individual; not with the system. There are some men unfitted by nature for aviation, just as there are others unfitted to be railway engineers.

CHAPTER XVI. RADICAL CHANGES BEING MADE.

Changes, many of them extremely radical in their nature, are continually being made by prominent aviators, and particularly those who have won the greatest amount of success. Wonderful as the results have been few of the aviators are really satisfied. Their successes have merely spurred them on to new endeavors, the ultimate end being the development of an absolutely perfect aircraft.

Among the men who have been thus experimenting are the Wright Brothers, who last year (1909) brought out a craft totally different as regards proportions and weight from the one used the preceding year. One marked result was a gain of about 3 1/2 miles an hour in speed.

Dimensions of 1908 Machine.

The 1908 model aeroplane was 40 by 29 feet over all. The carrying surfaces, that is, the two aerocurves, were 40 by 6 feet, having a parabolical curve of one in twelve. With about 70 square feet of surface in the rudders, the total surface given was about 550 square feet. The engine, which is the invention of the Wright brothers, weighed, approximately, 200 pounds, and gave about 25 horsepower at 1,400 revolutions per minute. The total weight of the aeroplane, exclusive of passenger, but inclusive of engine, was about 1,150 pounds. This result showed a lift of a fraction over 2 1/4 pounds to the square foot of carrying surface. The speed desired was 40 miles an hour, but the machine was found to make only a scant 39 miles an hour. The upright struts were about 7/8-inch thick, the skids, 2 1/2 by 1 1/4 inches thick.

Dimensions of 1909 Machine.

The 1909 aeroplane was built primarily for greater speed, and relatively heavier; to be less at the mercy of the wind. This result was obtained as follows: The aerocurves, or carrying surfaces, were reduced in dimensions from 40 by 6 feet to 36 by 5 1/2 feet, the curve remaining the same, one in twelve. The upright struts were cut from seven-eighths inch to five-eighths inch, and the skids from two and one-half by one and one-quarter to two and one-quarter by one and three-eighths inches. This result shows that there were some 81 square feet of carrying surface missing over that of last year's model. and some 25 pounds loss of weight. Relatively, though, the 1909 model aeroplane, while actually 25 pounds lighter, is really some 150 pounds heavier in the air than the 1908 model, owing to the lesser square feet of carrying surface.

Some of the Results Obtained.

Reducing the carrying surfaces from 6 to 5 1/2 feet gave two results—first, less carrying capacity; and, second, less head-on resistance, owing to the fact that the extent of the parabolic curve in the carrying surfaces was shortened. The "head-on" resistance is the retardance the aeroplane meets in passing through the air, and is counted in square feet. In the 1908 model the curve being one in twelve and 6 feet deep, gave 6 inches of head-on resistance. The plane being 40 feet spread, gave 6 inches by 40 feet, or 20 square feet of head-on resistance. Increasing this figure by a like amount for each plane, and adding approximately 10 square feet for struts, skids and wiring, we have a total of approximately, 50 square feet of surface for "head-on" resistance.

In the 1909 aeroplane, shortening the curve 6 inches at the parabolic end of the curve took off 1 inch of head-on resistance. Shortening the spread of the planes took off between 3 and 4 square feet of head-on resistance. Add to this the total of 7 square feet, less curve surface and about 1 square foot, less wire and woodwork resistance, and we have a grand total of, approximately, 12 square feet of less "head-on" resistance over the 1908 model.

Changes in Engine Action.

The engine used in 1909 was the same one used in 1908, though some minor changes were made as improvements; for instance, a make and break spark was used, and a nine-tooth, instead of a ten-tooth magneto gear-wheel was used. This increased the engine revolutions per minute from 1,200 to 1,400, and the propeller revolutions per minute from 350 to 371, giving a propeller thrust of, approximately, 170 foot pounds instead of 153, as was had last year.

More Speed and Same Capacity.

One unsatisfactory feature of the 1909 model over that of 1908, apparently, was the lack of inherent lateral stability. This was caused by the lesser surface and lesser extent of curvatures at the portions of the aeroplane which were warped. This defect did not show so plainly after Mr. Orville Wright had become fully proficient in the handling of the new machine, and with skillful management, the 1909 model aeroplane will be just as safe and secure as the other though it will take a little more practice to get that same degree of skill.

To sum up: The aeroplane used in 1909 was 25 pounds lighter, but really about 150 pounds heavier in the air, had less head-on resistance, and greater propeller thrust. The speed was increased from about 39 miles per hour to 42 1/2 miles per hour. The lifting capacity remained about the same, about 450 pounds capacity passenger-weight, with the 1908 machine. In this respect, the loss of carrying surface was compensated for by the increased speed.

During the first few flights it was plainly demonstrated that it would need the highest skill to properly handle the aeroplane, as first one end and then the other would dip and strike the ground, and either tear the canvas or slew the aeroplane around and break a skid.

Wrights Adopt Wheeled Gears.

In still another important respect the Wrights, so far as the output of one of their companies goes, have made a radical change. All the aeroplanes turned out by the Deutsch Wright Gesellschaft, according to the German publication, *Automobil-Welt*, will hereafter be equipped with wheeled running gears and tails. The plan of this new machine is shown in the illustration on page 145. The wheels are three in number, and are attached one to each of the two skids, just under the front edge of the planes, and one forward of these, attached to a cross-member. It is asserted that with these wheels the teaching of purchasers to operate the machines is much simplified, as the beginners can make short flights on their own account without using the starting derrick.

This is a big concession for the Wrights to make, as they have hitherto adhered stoutly to the skid gear. While it is true they do not control the German company producing their aeroplanes, yet the nature of their connection with the enterprise is such that it may be taken for granted no radical changes in construction would be made without their approval and consent.

Only Three Dangerous Rivals.

Official trials with the 1909 model smashed many records and leave the Wright brothers with only three dangerous rivals in the field, and with basic patents which cover the curve, warp and wingtip devices found on all the other makes of aeroplanes. These three rivals are the Curtiss and Voisin biplane type and the Bleriot monoplane pattern.

The Bleriot monoplane is probably the most dangerous rival, as this make of machine has a record of 54 miles per hour, has crossed the English channel, and has lifted two passengers besides the operator. The latest type of this machine only weighs 771.61 pounds complete, without passengers, and will lift a total passenger weight of 462.97 pounds, which is a lift of 5.21 pounds to the square foot. This is a better result than those published by the Wright brothers, the best noted being 4.25 pounds per square foot.

Other Aviators at Work.

The Wrights, however, are not alone in their efforts to promote the efficiency of the flying machine. Other competent inventive aviators, notably Curtiss, Voisin, Bleriot and Farman, are close after them. The Wrights, as stated, have a marked advantage in the possession of patents covering surface plane devices which have thus far been found indispensable in flying machine construction. Numerous law suits growing out of alleged infringements of these patents have been started, and others are threatened. What effect these actions will have in deterring aviators in general from proceeding with their experiments remains to be seen.

In the meantime the four men named—Curtiss, Voisin, Bleriot and Farman—are going ahead regardless of consequences, and the inventive genius of each is so strong that it is reasonable to expect some remarkable developments in the near future.

Smallest of Flying Machines.

To Santos Dumont must be given the credit of producing the smallest practical flying machine yet constructed. True, he has done nothing remarkable with it in the line of speed, but he has demonstrated the fact that a large supporting surface is not an essential feature.

This machine is named "La Demoiselle." It is a monoplane of the dihedral type, with a main plane on each side of the center. These main planes are of 18 foot spread, and nearly 6 1/2 feet in depth, giving approximately 115 feet of surface area. The total weight is 242 pounds, which is 358 pounds less than any other machine which has been successfully used. The total depth from front to rear is 26 feet.

The framework is of bamboo, strengthened and held taut with wire guys.

Have One Rule in Mind.

In this struggle for mastery in flying machine efficiency all the contestants keep one rule in mind, and this is:

"The carrying capacity of an aeroplane is governed by the peripheral curve of its carrying surfaces, plus the speed; and the speed is governed by the thrust of the propellers, less the 'head-on' resistance."

Their ideas as to the proper means of approaching the proposition may, and undoubtedly are, at variance, but the one rule in solving the problem of obtaining the greatest carrying capacity combined with the greatest speed, obtains in all instances.

CHAPTER XVII. SOME OF THE NEW DESIGNS.

Spurred on by the success attained by the more experienced and better known aviators numerous inventors of lesser fame are almost daily producing practical flying machines varying radically in construction from those now in general use.

One of these comparatively new designs is the Van Anden biplane, made by Frank Van Anden of Islip, Long Island, a member of the New York Aeronautic Society. While his machine is wholly experimental, many successful short flights were made with it last fall (1909). One flight, made October 19th, 1909, is of particular interest as showing the practicability of an automatic stabilizing device installed by the inventor. The machine was caught in a sudden severe gust of wind and keeled over, but almost immediately righted itself, thus demonstrating in a most satisfactory manner the value of one new attachment.

Features of Van Anden Model.

In size the surfaces of the main biplane are 26 feet in spread, and 4 feet in depth from front to rear. The upper and lower planes are 4 feet apart. Silkolene coated with varnish is used for the coverings. Ribs (spruce) are curved one inch to the foot, the deepest part of the curve (4 inches) being one foot back from the front edge of the horizontal beam. Struts (also of spruce, as is all the framework) are elliptical in shape. The main beams are in three sections, nearly half round in form, and joined by metal sleeves.

There is a two-surface horizontal rudder, 2x2x4 feet, in front. This is pivoted at its lateral center 8 feet from the front edge of the main planes. In the rear is another two-surface horizontal rudder 2x2x2 1/2 feet, pivoted in the same manner as the front one, 15 feet from the rear edges of the main planes.

Hinged to the rear central strut of the rear rudder is a vertical rudder 2 feet high by 3 feet in length.

The Method of Control.

In the operation of these rudders—both front and rear—and the elevation and depression of the main planes, the Curtiss system is employed. Pushing the steering-wheel post outward depresses the front edges of the planes, and brings the machine downward; pulling the steering-wheel post inward elevates the front edges of the planes and causes the machine to ascend.

Turning the steering wheel itself to the right swings the tail rudder to the left, and the machine, obeying this like a boat, turns in the same direction as the wheel is turned. By like cause turning the wheel to the left turns the machine to the left.

Automatic Control of Wings.

There are two wing tips, each of 6 feet spread (length) and 2 feet from front to rear. These are hinged half way between the main surfaces to the two outermost rear struts. Cables run from these to an automatic device working with power from the engine, which automatically operates the tips with the tilting of the machine. Normally the wing tips are held horizontal by stiff springs introduced in the cables outside of the device.

It was the successful working of this device which righted the Van Anden craft when it was overturned in the squall of October 19th, 1909. Previous to that occurrence Mr. Van Anden had looked upon the device as purely experimental, and had admitted that he had grave uncertainty as to how it would operate in time of emergency. He is now quoted as being thoroughly satisfied with its practicability. It is this automatic device which gives the Van Anden machine at least one distinctively new feature.

While on this subject it will not be amiss to add that Mr. Curtiss does not look kindly on automatic control. "I would rather trust to my own action than that of a machine," he says. This is undoubtedly good logic so far as Mr. Curtiss is concerned, but all aviators are not so cool-headed and resourceful.

Motive Power of Van Anden.

A 50-horsepower "H-F" water cooled motor drives a laminated wood propeller 6 feet in diameter, with a 17 degree pitch at the extremities, increasing toward the hub. The rear end of the motor is about 6 inches back from the rear transverse beam and the engine shaft is in a direct line with the axes of the two horizontal rudders. An R. I. V. ball bearing carries the shaft at this point. Flying, the motor turns at about 800 revolutions per minute, delivering 180 pounds pull. A test of the motor running at 1,200 showed a pull of 250 pounds on the scales.

Still Another New Aeroplane.

Another new aeroplane is that produced by A. M. Herring (an old-timer) and W. S. Burgess, under the name of the Herring-Burgess. This is also equipped with an automatic stability device for maintaining the balance transversely. The curvature of the planes is also laid out on new lines. That this new plan is effective is evidenced by the fact that the machine has been elevated to an altitude of 40 feet by using one-half the power of the 30-horsepower motor.

The system of rudder and elevation control is very simple. The aviator sits in front of the lower plane, and extending his arms, grasps two supports which extend down diagonally in front. On the under side of these supports just beneath his fingers are the controls which operate the vertical rudder, in the rear. Thus, if he wishes to turn to the right, he presses the control under the fingers of his right hand; if to the left, that under the fingers of his left hand. The elevating rudder is operated by the aviator's right foot, the control being placed on a foot-rest.

Motor Is Extremely Light.

Not the least notable feature of the craft is its motor. Although developing, under load, 30-horsepower, or that of an ordinary automobile, it weighs, complete, hardly 100 pounds. Having occasion to move it a little distance for inspection, Mr. Burgess picked it up and walked off with it — cylinders, pistons, crankcase and all, even the magneto, being attached. There are not many 30-horsepower engines which can be so handled. Everything about it is reduced to its lowest terms of simplicity, and hence, of weight. A single camshaft operates not only all of the inlet and exhaust valves, but the magneto and gear water pump, as well. The motor is placed directly behind the operator, and the propeller is directly mounted on the crankshaft.

This weight of less than 100 pounds, it must be remembered, is not for the motor alone; it includes the entire power plant equipment.

The "thrust" of the propeller is also extraordinary, being between 250 and 260 pounds. The force of the wind displacement is strong enough to knock down a good-sized boy as one youngster ascertained when he got behind the propeller as it was being tested. He was not only knocked down but driven for some distance away from the machine. The propeller has four blades which are but little wider than a lath.

Machine Built by Students.

Students at the University of Pennsylvania, headed by Laurence J. Lesh, a protege of Octave Chanute, have constructed a practical aeroplane of ordinary maximum size, in which is incorporated many new ideas. The most unique of these is to be found in the steering gear, and the provision made for the accommodation of a pupil while taking lessons under an experienced aviator.

Immediately back of the aviator is an extra seat and an extra steering wheel which works in tandem style with the front wheel. By this arrangement a beginner may be easily and quickly taught to have perfect control of the machine. These tandem wheels are also handy for passengers who may wish to operate the car independently of one another, it being understood, of course, that there will be no conflict of action.

Frame Size and Engine Power.

The frame has 36 feet spread and measures 35 feet from the front edge to the end of the tail in the rear. It is equipped with two rear propellers operated by a Ramsey 8-cylinder motor of 50 horsepower, placed horizontally across the lower plane, with the crank shaft running clear through the engine.

The "Pennsylvania I" is the first two-propeller biplane chainless car, this scheme having been adopted in order to avoid the crossing of chains. The lateral control is by a new invention by Octave Chanute and Laurence J. Lesh, for which Lesh is now applying for a patent. The device was worked out before the Wright brothers' suit was begun, and is said to be superior to the Wright warping or the Curtiss ailerons. The landing device is also new in design. This aeroplane will weigh about 1,500 pounds, and will carry fuel for a flight of 150 miles, and it is expected to attain a speed of at least 45 miles an hour.

There are others, lots of them, too numerous in fact to admit of mention in a book of this size.

CHAPTER XVIII. DEMAND FOR FLYING MACHINES.

As a commercial proposition the manufacture and sale of motor-equipped aeroplanes is making much more rapid advance than at first obtained in the similar handling of the automobile. Great, and even phenomenal, as was the commercial development of the motor car, that of the flying machine is even greater. This is a startling statement, but it is fully warranted by the facts.

It is barely more than a year ago (1909) that attention was seriously attracted to the motor-equipped aeroplane as a vehicle possible of manipulation by others than professional aviators. Up to that time such actual flights as were made were almost exclusively with the sole purpose of demonstrating the practicability of the machine, and the merits of the ideas as to shape, engine power, etc., of the various producers.

Results of Bleriot's Daring.

It was not until Bleriot flew across the straits of Dover on July 25th, 1909, that the general public awoke to a full realization of the fact that it was possible for others than professional aviators to indulge in aviation. Bleriot's feat was accepted as proof that at last an absolutely new means of sport, pleasure and research, had been practically developed, and was within the reach of all who had the inclination, nerve and financial means to adopt it.

From this event may be dated the birth of the modern flying machine into the world of business. The automobile was taken up by the general public from the very start because it was a proposition comparatively easy of demonstration. There was nothing mysterious or uncanny in the fact that a wheeled vehicle could be propelled on solid, substantial roads by means of engine power. And yet it took (comparatively speaking) a long time to really popularize the motor car.

Wonderful Results in a Year.

Men of large financial means engaged in the manufacture of automobiles, and expended fortunes in attracting public attention to them through the medium of advertisements, speed and road contests, etc. By these means a mammoth business has been built up, but bringing this business to its present proportions required years of patient industry and indomitable pluck.

At this writing, less than a year from the day when Bleriot crossed the channel, the actual sales of flying machines outnumber the actual sales of automobiles in the first year of their commercial development. This may appear incredible, but it is a fact as statistics will show.

In this connection we should take into consideration the fact that up to a year ago there was no serious intention of putting flying machines on the market; no preparations had been made to produce them on a commercial scale; no money had been expended in advertisements with a view to selling them.

Some of the Actual Results.

Today flying machines are being produced on a commercial basis, and there is a big demand for them. The people making them are overcrowded with orders. Some of the producers are already making arrangements to enlarge their plants and advertise their product for sale the same as is being done with automobiles, while a number of flying machine motor makers are already promoting the sale of their wares in this way.

Here are a few actual figures of flying machine sales made by the more prominent producers since July 25th, 1909.

Santos Dumont, 90 machines; Bleriot, 200; Farman, 130; Clemenceau-Wright, 80; Voisin, 100; Antoinette, 100. Many of these orders have been filled by delivery of the machines, and in others the construction work is under way.

The foregoing are all of foreign make. In this country Curtiss and the Wrights are engaged in similar work, but no actual figures of their output are obtainable.

Larger Plants Are Necessary.

And this situation exists despite the fact that none of the producers are really equipped with adequate plants for turning out their machines on a modern, business-like basis. The demand was so sudden and unexpected that it found them poorly prepared to meet it. This, however, is now being remedied by the erection of special plants, the enlargement of others, and the introduction of new machinery and other labor-saving conveniences.

Companies, with large capitalization, to engage in the exclusive production of airships are being organized in many parts of the world. One notable instance of this nature is worth quoting as illustrative of the manner in which the production of flying machines is being commercialized. This is the formation at Frankfort, Germany, of the Flugmaschine Wright, G. m. b. H., with a capital of $119,000, the Krupps, of Essen, being interested.

Prices at Which Machines Sell.

This wonderful demand from the public has come notwithstanding the fact that the machines, owing to lack of facilities for wholesale production, are far from being cheap. Such definite quotations as are made are on the following basis:

Santos Dumont—List price $1,000, but owing to the rush of orders agents are readily getting from $1,300 to $1,500. This is the smallest machine made.

Bleriot—List price $2,500. This is for the cross-channel type, with Anzani motor.

Antoinette—List price from $4,000 to $5,000, according to size.

Wright—List price $5,600.

Curtiss—List price $5,000.

There is, however, no stability in prices as purchasers are almost invariably ready to pay a considerable premium to facilitate delivery.

The motor is the most expensive part of the flying machine. Motor prices range from $500 to $2,000, this latter amount being asked for the Curtiss engine.

Systematic Instruction of Amateurs.

In addition to the production of flying machines many of the experienced aviators are making a business of the instruction of amateurs. Curtiss and the Wrights in this country have a number of pupils, as have also the prominent foreigners. Schools of instruction are being opened in various parts of the world, not alone as private money-making ventures, but in connection with public educational institutions. One of these latter is to be found at the University of Barcelona, Spain.

The flying machine agent, the man who handles the machines on a commission, has also become a known quantity, and will soon be as numerous as his brother of the automobile. The sign "John Bird, agent for Skimmer's Flying Machine," is no longer a curiosity.

Yes, the Airship Is Here.

From all of which we may well infer that the flying machine in practical form has arrived, and that it is here to stay. It is no exaggeration to say that the time is close at hand when people will keep flying machines just as they now keep automobiles, and that pleasure jaunts will be fully as numerous and popular. With the important item of practicability fully demonstrated, "Come, take a trip in my airship," will have more real significance than now attaches to the vapid warblings of the vaudeville vocalist.

As a further evidence that the airship is really here, and that its presence is recognized in a business way, the action of life and accident insurance companies is interesting. Some of them are reconstructing their policies so as to include a special waiver of insurance by aviators. Anything which compels these great corporations to modify their policies cannot be looked upon as a mere curiosity or toy.

It is some consolation to know that the movement in this direction is not thus far widespread. Moreover it is more than probable that the competition for business will eventually induce the companies to act more liberally toward aviators, especially as the art of aviation advances.

CHAPTER XIX. LAW OF THE AIRSHIP.

Successful aviation has evoked some peculiar things in the way of legal action and interpretation of the law.

It is well understood that a man's property cannot be used without his consent. This is an old established principle in common law which holds good today.

The limits of a man's property lines, however, have not been so well understood by laymen. According to eminent legal authorities such as Blackstone, Littleton and Coke, the "fathers of the law," the owner of realty also holds title above and below the surface, and this theory is generally accepted without question by the courts.

Rights of Property Owners.

In other words the owner of realty also owns the sky above it without limit as to distance. He can dig as deep into his land, or go as high into the air as he desires, provided he does not trespass upon or injure similar rights of others.

The owner of realty may resist by force, all other means having failed, any trespass upon, or invasion of his property. Other people, for instance, may not enter upon it, or over or under it, without his express permission and consent. There is only one exception, and this is in the case of public utility corporations such as railways which, under the law of eminent domain, may condemn a right of way across the property of an obstinate owner who declines to accept a fair price for the privilege.

Privilege Sharply Confined.

The law of eminent domain may be taken advantage of only by corporations which are engaged in serving the public. It is based upon the principle that the advancement and improvement of a community is of more importance and carries with it more rights than the interests of the individual owner. But even in cases where the right of eminent domain is exercised there can be no confiscation of the individual's property.

Exercising the right of eminent domain is merely obtaining by public purchase what is held to be essential to the public good, and which cannot be secured by private purchase. When eminent domain proceedings are resorted to the court appoints appraisers who determine upon the value of the property wanted, and this value (in money) is paid to the owner.

How It Affects Aviation.

It should be kept in mind that this privilege of the "right of eminent domain" is accorded only to corporations which are engaged in serving the public. Individuals cannot take advantage of it. Thus far all aviation has been conducted by individuals; there are no flying machine or airship corporations regularly engaged in the transportation of passengers, mails or freight.

This leads up to the question "What would happen if realty owners generally, or in any considerable numbers, should prohibit the navigation of the air above their holdings?" It is idle to say such a possibility is ridiculous—it is already an actuality in a few individual instances.

One property owner in New Jersey, a justice of the peace, maintains a large sign on the roof of his house warning aviators that they must not trespass upon his domain. That he is acting well within his rights in doing this is conceded by legal authorities.

Hard to Catch Offenders.

But, suppose the alleged trespass is committed, what is the property owner going to do about it? He must first catch the trespasser and this would be a pretty hard job. He certainly could not overtake him, unless he kept a racing aeroplane for this special purpose. It would be equally difficult to identify the offender after the offense had been committed, even if he were located, as aeroplanes carry no license numbers.

Allowing that the offender should be caught the only recourse of the realty owner is an action for damages. He may prevent the commission of the offense by force if necessary, but after it is committed he can only sue for damages. And in doing this he would have a lot of trouble.

Points to Be Proven.

One of the first things the plaintiff would be called upon to prove would be the elevation of the machine. If it were reasonably close to the ground there would, of course, be grave risk of damage to fences, shrubbery, and other property, and the court would be justified in holding it to be a nuisance that should be suppressed.

If, on the other hand; the machine was well up in the air, but going slowly, or hovering over the plaintiff's property, the court might be inclined to rule that it could not possibly be a nuisance, but right here the court would be in serious embarrassment. By deciding that it was not a nuisance he would virtually override the law against invasion of a man's property without his consent regardless of the nature of the invasion. By the same decision he would also say in effect that, if one flying machine could do this a dozen or more would have equal right to do the same thing. While one machine hovering over a certain piece of property may be no actual nuisance a dozen or more in the same position could hardly be excused.

Difficult to Fix Damages.

Such a condition would tend to greatly increase the risk of accident, either through collision, or by the carelessness of the aviators in dropping articles which might cause damages to the people or property below. In such a case it would undoubtedly be a nuisance, and in addition to a fine, the offender would also be liable for the damages.

Taking it for granted that no actual damage is done, and the owner merely sues on account of the invasion of his property, how is the amount of compensation to be fixed upon? The owner has lost nothing; no part of his possessions has been taken away; nothing has been injured or destroyed; everything is left in exactly the same condition as before the invasion. And yet, if the law is strictly interpreted, the offender is liable.

Right of Way for Airships.

Somebody has suggested the organization of flying-machine corporations as common carriers, which would give them the right of eminent domain with power to condemn a right of way. But what would they condemn? There is nothing tangible in the air. Railways in condemning a right of way specify tangible property (realty) within certain limits. How would an aviator designate any particular right of way through the air a certain number of feet in width, and a certain distance from the ground?

And yet, should the higher courts hold to the letter of the law and decide that aviators have no right to navigate their craft over private property, something will have to be done to get them out of the dilemma, as aviation is too far advanced to be discarded. Fortunately there is little prospect of any widespread antagonism among property owners so long as aviators refrain from making nuisances of themselves.

Possible Solution Offered.

One possible solution is offered and that is to confine the path of airships to the public highways so that nobody's property rights would be invaded. In addition, as a matter of promoting safety for both operators and those who may happen to be beneath the airships as they pass over a course, adoption of the French rules are suggested. These are as follows:

Aeroplanes, when passing, must keep to the right, and pass at a distance of at least 150 feet. They are free from this rule when flying at altitudes of more than 100 feet. Every machine when flying at night or during foggy weather must carry a green light on the right, and a red light on the left, and a white headlight on the front.

These are sensible rules, but may be improved upon by the addition of a signal system of some kind, either horn, whistle or bell.

Responsibility of Aviators.

Mr. Jay Carver Bossard, in recent numbers of *Fly*, brings out some curious and interesting legal points in connection with aviation, among which are the following:

"Private parties who possess aerial craft, and desire to operate the same in aerial territory other than their own, must obtain from land owners special permission to do so, such permission to be granted only by agreement, founded upon a valid consideration. Otherwise, passing over another's land will in each instance amount to a trespass.

"Leaving this highly technical side of the question, let us turn to another view: the criminal and tort liability of owners and operators to airship passengers. If A invites B to make an ascension with him in his machine, and B, knowing that A is merely an enthusiastic amateur and far from being an expert, accepts and is through A's innocent negligence injured, he has no grounds for recovery. But if A contracts with B, to transport him from one place to another, for a consideration, and B is injured by the poor piloting of A, A would be liable to B for damages which would result. Now in order to safeguard such people as B, curious to the point of recklessness, the law will have to require all airship operators to have a license, and to secure this license airship pilots will have to meet certain requirements. Here again is a question. Who is going to say whether an applicant is competent to pilot a balloon or airship?

Fine for an Aeronaut.

"An aeroplane while maneuvering is suddenly caught by a treacherous gale and swept to the ground. A crowd of people hasten over to see if the aeronaut is injured, and in doing so trample over Tax-payer Smith's garden, much to the detriment of his growing vegetables and flowers. Who is liable for the damages? Queer as it may seem, a case very similar to this was decided in 1823, in the New York supreme court, and it was held that the aeronaut was liable upon the following grounds: 'To render one man liable in trespass for the acts of others, it must appear either that they acted in concert, or that the act of the one, ordinarily and naturally produced the acts of the others, Ascending in a balloon is not an unlawful act, but it is certain that the aeronaut has no control over its motion horizontally, but is at the sport of the wind, and is to descend when and how he can. His reaching the earth is a matter of hazard. If his descent would according to the circumstances draw a crowd of people around him, either out of curiosity, or for the purpose of rescuing him from a perilous situation, all this he ought to have foreseen, and must be responsible for.'

Air Not Really Free.

"The general belief among people is, that the air is free. Not only free to breathe and enjoy, but free to travel in, and that no one has any definite jurisdiction over, or in any part of it. Now suppose this were made a legal doctrine. Would a murder perpetrated above the clouds have to go unpunished? Undoubtedly. For felonies committed upon the high seas ample provision is made for their punishment, but new provisions will have to be made for crimes committed in the air.

Relations of Owner and Employee.

"It is a general rule of law that a master is bound to provide reasonably safe tools, appliances and machines for his servant. How this rule is going to be applied in cases of aeroplanes, remains to be seen. The aeroplane owner who hires a professional aeronaut, that is, one who has qualified as an expert, owes him very little legal duty to supply him with a perfect aeroplane. The expert is supposed to know as much regarding the machine as the owner, if not more, and his acceptance of his position relieves the owner from liability. When the owner hires an amateur aeronaut to run the aeroplane, and teaches him how to manipulate it, even though the prescribed manner of manipulation will make flight safe, nevertheless if the machine is visibly defective, or known to be so, any injury which results to the aeronaut the owner is liable for.

As to Aeroplane Contracts.

"At the present time there are many orders being placed with aeroplane manufacturing companies. There are some unique questions to be raised here under the law of contract. It is an elementary principle of law that no one can be compelled to complete a contract which in itself is impossible to perform. For instance, a contract to row a boat across the Atlantic in two weeks, for a consideration, could never be enforced because it is within judicial knowledge that such an undertaking is beyond human power. Again, contracts formed for the doing of acts contrary to nature are never enforcible, and here is where our difficulty comes in. Is it possible to build a machine or species of craft which will transport a person or goods through the air? The courts know that balloons are practical; that is, they know that a bag filled with gas has a lifting power and can move through the air at an appreciable height. Therefore, a contract to transport a person in such manner is a good contract, and the conditions being favorable could undoubtedly be enforced. But the passengers' right of action for injury would be very limited.

No Redress for Purchasers.

"In the case of giving warranties on aeroplanes, we have yet to see just what a court is going to say. It is easy enough for a manufacturer to guarantee to build a machine of certain dimensions and according to certain specifications, but when he inserts a clause in the contract to the effect that the machine will raise itself from the surface of the earth, defy the laws of gravity, and soar in the heavens at the will of the aviator, he is to say the least contracting to perform a miracle.

"Until aeroplanes have been made and accepted as practical, no court will force a manufacturer to turn out a machine guaranteed to fly. So purchasers can well remember that if their machines refuse to fly they have no redress against the maker, for he can always say, 'The industry is still in its experimental stage.' In contracting for an engine no builder will guarantee that the particular engine will successfully operate the aeroplane. In fact he could never be forced to live up to such an agreement, should he agree to a stipulation of that sort. The best any engine maker will guarantee is to build an engine according to specifications."

CHAPTER XX. SOARING FLIGHT.

By Octave Chanute.

5 There is a wonderful performance daily exhibited in southern climes and occasionally seen in northerly latitudes in summer, which has never been thoroughly explained. It is the soaring or sailing flight of certain varieties of large birds who transport themselves on rigid, unflapping wings in any desired direction; who in winds of 6 to 20 miles per hour, circle, rise, advance, return and remain aloft for hours without a beat of wing, save for getting under way or convenience in various maneuvers. They appear to obtain from the wind alone all the necessary energy, even to advancing dead against that wind. This feat is so much opposed to our general ideas of physics that those who have not seen it sometimes deny its actuality, and those who have only occasionally witnessed it subsequently doubt the evidence of their own eyes. Others, who have seen the exceptional performances, speculate on various explanations, but the majority give it up as a sort of "negative gravity."

Soaring Power of Birds.

The writer of this paper published in the "Aeronautical Annual" for 1896 and 1897 an article upon the sailing flight of birds, in which he gave a list of the authors who had described such flight or had advanced theories for its explanation, and he passed these in review. He also described his own observations and submitted some computations to account for the observed facts. These computations were correct as far as they went, but they were scanty. It was, for instance, shown convincingly by analysis that a gull weighing 2.188 pounds, with a total supporting surface of 2.015 square feet, a maximum body cross-section of 0.126 square feet and a maximum cross-section of wing edges of 0.098 square feet, patrolling on rigid wings (soaring) on the weather side of a steamer and maintaining an upward angle or attitude of 5 degrees to 7 degrees above the horizon, in a wind blowing 12.78 miles an hour, which was deflected upward 10 degrees to 20 degrees by the side of the steamer (these all being carefully observed facts), was perfectly sustained at its own "relative speed" of 17.88 miles per hour and extracted from the upward trend of the wind sufficient energy to overcome all the resistances, this energy amounting to 6.44 foot-pounds per second.

Great Power of Gulls.

It was shown that the same bird in flapping flight in calm air, with an attitude or incidence of 3 degrees to 5 degrees above the horizon and a speed of 20.4 miles an hour was well sustained and expended 5.88 foot-pounds per second, this being at the rate of 204 pounds sustained per horsepower. It was stated also that a gull in its observed maneuvers, rising up from a pile head on unflapping wings, then plunging forward against the wind and subsequently rising higher than his starting point, must either time his ascents and descents exactly with the variations in wind velocities, or must meet a wind billow rotating on a horizontal axis and come to a poise on its crest, thus availing of an ascending trend.

But the observations failed to demonstrate that the variations of the wind gusts and the movements of the bird were absolutely synchronous, and it was conjectured that the peculiar shape of the soaring wing of certain birds, as differentiated from the flapping wing, might, when experimented upon, hereafter account for the performance.

Mystery to be Explained.

These computations, however satisfactory they were for the speed of winds observed, failed to account for the observed spiral soaring of buzzards in very light winds and the writer was compelled to confess: "Now, this spiral soaring in steady breezes of 5 to 10 miles per hour which are apparently horizontal, and through which the bird maintains an average speed of about 20 miles an hour, is the mystery to be explained. It is not accounted for, quantitatively, by any of the theories which have been advanced, and it is the one performance which has led some observers to claim that it was done through 'aspiration.' i, e., that a bird acted upon by a current, actually drew forward into that current against its exact direction of motion."

Buzzards Soar in Dead Calm.

A still greater mystery was propounded by the few observers who asserted that they had seen buzzards soaring in a dead calm, maintaining their elevation and their speed. Among these observers was Mr. E. C. Huffaker, at one time assistant experimenter for Professor Langley. The writer believed and said then that he must in some way have been mistaken, yet, to satisfy himself, he paid several visits to Mr. Huffaker, in Eastern Tennessee and took along his anemometer. He saw quite a number of buzzards sailing at a height of 75 to 100 feet in breezes measuring 5 or 6 miles an hour at the surface of the ground, and once he saw one buzzard soaring apparently in a dead calm.

The writer was fairly baffled. The bird was not simply gliding, utilizing gravity or acquired momentum, he was actually circling horizontally in defiance of physics and mathematics. It took two years and a whole series of further observations to bring those two sciences into accord with the facts.

Results of Close Observations.

Curiously enough the key to the performance of circling in a light wind or a dead calm was not found through the usual way of gathering human knowledge, i. e., through observations and experiment. These had failed because I did not know what to look for. The mystery was, in fact, solved by an eclectic process of conjecture and computation, but once these computations indicated what observations should be made, the results gave at once the reasons for the circling of the birds, for their then observed attitude, and for the necessity of an independent initial sustaining speed before soaring began. Both Mr. Huffaker and myself verified the data many times and I made the computations.

These observations disclosed several facts:

1st.—That winds blowing five to seventeen miles per hour frequently had rising trends of 10 degrees to 15 degrees, and that upon occasions when there seemed to be absolutely no wind, there was often nevertheless a local rising of the air estimated at a rate of four to eight miles or more per hour. This was ascertained by watching thistledown, and rising fogs alongside of trees or hills of known height. Everyone will readily realize that when walking at the rate of four to eight miles an hour in a dead calm the "relative wind" is quite inappreciable to the senses and that such a rising air would not be noticed.

2nd.—That the buzzard, sailing in an apparently dead horizontal calm, progressed at speeds of fifteen to eighteen miles per hour, as measured by his shadow on the ground. It was thought that the air was then possibly rising 8.8 feet per second, or six miles per hour.

3rd.—That when soaring in very light winds the angle of incidence of the buzzards was negative to the horizon—i. e., that when seen coming toward the eye, the afternoon light shone on the back instead of on the breast, as would have been the case had the angle been inclined above the horizon.

4th.—That the sailing performance only occurred after the bird had acquired an initial velocity of at least fifteen or eighteen miles per hour, either by industrious flapping or by descending from a perch.

An Interesting Experiment.

5th.—That the whole resistance of a stuffed buzzard, at a negative angle of 3 degrees in a current of air of 15.52 miles per hour, was 0.27 pounds. This test was kindly made for the writer by Professor A. F. Zahm in the "wind tunnel" of the Catholic University at Washington, D. C., who, moreover, stated that the resistance of a live bird might be less, as the dried plumage could not be made to lie smooth.

This particular buzzard weighed in life 4.25 pounds, the area of his wings and body was 4.57 square feet, the maximum cross-section of his body was 0.110 square feet, and that of his wing edges when fully extended was 0.244 square feet.

With these data, it became surprisingly easy to compute the performance with the coefficients of Lilienthal for various angles of incidence and to demonstrate how this buzzard could soar horizontally in a dead horizontal calm, provided that it was not a vertical calm, and that the air was rising at the rate of four or six miles per hour, the lowest observed, and quite inappreciable without actual measuring.

Some Data on Bird Power.

The most difficult case is purposely selected. For if we assume that the bird has previously acquired an initial minimum speed of seventeen miles an hour (24.93 feet per second, nearly the lowest measured), and that the air was rising vertically six miles an hour (8.80 feet per second), then we have as the trend of the "relative wind" encountered:

$$6 - = 0.353, \text{ or the tangent of } 19 \text{ degrees } 26'. 17$$

which brings the case into the category of rising wind effects. But the bird was observed to have a negative angle to the horizon of about 3 degrees, as near as could be guessed, so that his angle of incidence to the "relative wind" was reduced to 16 degrees 26'.

The relative speed of his soaring was therefore:

Velocity = square root of (17 squared + 6 squared) = 18.03 miles per hour.

At this speed, using the Langley co-efficient recently practically confirmed by the accurate experiments of Mr. Eiffel, the air pressure would be:

18.03 squared X 0.00327 = 1.063 pounds per square foot.

If we apply Lilienthal's co-efficients for an angle of 6 degrees 26', we have for the force in action:

> Normal: 4.57 X 1.063 X 0.912 = 4.42 pounds. Tangential: 4.57 X 1.063 X 0.074 = - 0.359 pounds, which latter, being negative, is a propelling force.

Results Astonish Scientists.

Thus we have a bird weighing 4.25 pounds not only thoroughly supported, but impelled forward by a force of 0.359 pounds, at seventeen miles per hour, while the experiments of Professor A. F. Zahm showed that the resistance at 15.52 miles per hour was only 0.27 pounds,

> 17 squared or 0.27 X — — —- = 0.324 pounds, at seventeen miles an 15.52 squared hour.

These are astonishing results from the data obtained, and they lead to the inquiry whether the energy of the rising air is sufficient to make up the losses which occur by reason of the resistance and friction of the bird's body and wings, which, being rounded, do not encounter air pressures in proportion to their maximum cross-section.

We have no accurate data upon the co-efficients to apply and estimates made by myself proved to be much smaller than the 0.27 pounds resistance measured by Professor Zahm, so that we will figure with the latter as modified. As the speed is seventeen miles per hour, or 24.93 feet per second, we have for the work:

Work done, 0.324 X 24.93 = 8.07 foot pounds per second.

Endorsed by Prof. Marvin.

Corresponding energy of rising air is not sufficient at four miles per hour. This amounts to but 2.10 foot pounds per second, but if we assume that the air was rising at the rate of seven miles per hour (10.26 feet per second), at which the pressure with the Langley coefficient would be 0.16 pounds per square foot, we have on 4.57 square feet for energy of rising air: 4.57 X 0.16 X 10.26 = 7.50 foot pounds per second, which is seen to be still a little too small, but well within the limits of error, in view of the hollow shape of the bird's wings, which receive greater pressure than the flat planes experimented upon by Langley.

These computations were chiefly made in January, 1899, and were communicated to a few friends, who found no fallacy in them, but thought that few aviators would understand them if published. They were then submitted to Professor C. F. Marvin of the Weather Bureau, who is well known as a skillful physicist and mathematician. He wrote that they were, theoretically, entirely sound and quantitatively, probably, as accurate as the present state of the measurements of wind pressures permitted. The writer determined, however, to withhold publication until the feat of soaring flight had been performed by man, partly because he believed that, to ensure safety, it would be necessary that the machine should be equipped with a motor in order to supplement any deficiency in wind force.

Conditions Unfavorable for Wrights.

The feat would have been attempted in 1902 by Wright brothers if the local circumstances had been more favorable. They were experimenting on "Kill Devil Hill," near Kitty Hawk, N. C. This sand hill, about 100 feet high, is bordered by a smooth beach on the side whence come the sea breezes, but has marshy ground at the back. Wright brothers were apprehensive that if they rose on the ascending current of air at the front and began to circle like the birds, they might be carried by the descending current past the back of the hill and land in the marsh. Their gliding machine offered no greater head resistance in proportion than the buzzard, and their gliding angles of descent are practically as favorable, but the birds performed higher up in the air than they.

Langley's Idea of Aviation.

Professor Langley said in concluding his paper upon "The Internal Work of the Wind":

"The final application of these principles to the art of aerodromics seems, then, to be, that while it is not likely that the perfected aerodrome will ever be able to dispense altogether with the ability to rely at intervals on some internal source of power, it will not be indispensable that this aerodrome of the future shall, in order to go any distance—even to circumnavigate the globe without alighting—need to carry a weight of fuel which would enable it to perform this journey under conditions analogous to those of a steamship, but that the fuel and weight need only be such as to enable it to take care of itself in exceptional moments of calm."

Now that dynamic flying machines have been evolved and are being brought under control, it seems to be worth while to make these computations and the succeeding explanations known, so that some bold man will attempt the feat of soaring like a bird. The theory underlying the performance in a rising wind is not new, it has been suggested by Penaud and others, but it has attracted little attention because the exact data and the maneuvers required were not known and the feat had not yet been performed by a man. The puzzle has always been to account for the observed act in very light winds, and it is hoped that by the present selection of the most difficult case to explain—i. e., the soaring in a dead horizontal calm—somebody will attempt the exploit.

Requisites for Soaring Flights.

The following are deemed to be the requisites and maneuvers to master the secrets of soaring flight:

1st—Develop a dynamic flying machine weighing about one pound per square foot of area, with stable equilibrium and under perfect control, capable of gliding by gravity at angles of one in ten (5 3/4 degrees) in still air.

2nd.—Select locations where soaring birds abound and occasions where rising trends of gentle winds are frequent and to be relied on.

3rd.—Obtain an initial velocity of at least 25 feet per second before attempting to soar.

4th.—So locate the center of gravity that the apparatus shall assume a negative angle, fore and aft, of about 3 degrees.

Calculations show, however, that sufficient propelling force may still exist at 0 degrees, but disappears entirely at +4 degrees.

5th.—Circle like the bird. Simultaneously with the steering, incline the apparatus to the side toward which it is desired to turn, so that the centrifugal force shall be balanced by the centripetal force. The amount of the required inclination depends upon the speed and on the radius of the circle swept over.

6th.—Rise spirally like the bird. Steer with the horizontal rudder, so as to descend slightly when going with the wind and to ascend when going against the wind. The bird circles over one spot because the rising trends of wind are generally confined to small areas or local chimneys, as pointed out by Sir H. Maxim and others.

7th.—Once altitude is gained, progress may be made in any direction by gliding downward by gravity.

The bird's flying apparatus and skill are as yet infinitely superior to those of man, but there are indications that within a few years the latter may evolve more accurately proportioned apparatus and obtain absolute control over it.

It is hoped, therefore, that if there be found no radical error in the above computations, they will carry the conviction that soaring flight is not inaccessible to man, as it promises great economies of motive power in favorable localities of rising winds.

The writer will be grateful to experts who may point out any mistake committed in data or calculations, and will furnish additional information to any aviator who may wish to attempt the feat of soaring.

CHAPTER XXI. FLYING MACHINES VS. BALLOONS.

While wonderful success has attended the development of the dirigible (steerable) balloon the most ardent advocates of this form of aerial navigation admit that it has serious drawbacks. Some of these may be described as follows:

Expense and Other Items.

Great Initial Expense.—The modern dirigible balloon costs a fortune. The Zeppelin, for instance, costs more than $100,000 (these are official figures).

Expense of Inflation.—Gas evaporates rapidly, and a balloon must be re-inflated, or partially re-inflated, every time it is used. The Zeppelin holds 460,000 cubic feet of gas which, even at $1 per thousand, would cost $460.

Difficulty of Obtaining Gas.—If a balloon suddenly becomes deflated, by accident or atmospheric conditions, far from a source of gas supply, it is practically worthless. Gas must be piped to it, or the balloon carted to the gas house—an expensive proceeding in either event.

Lack of Speed and Control.

Lack of Speed.—Under the most favorable conditions the maximum speed of a balloon is 30 miles an hour. Its great bulk makes the high speed attained by flying machines impossible.

Difficulty of Control.—While the modern dirigible balloon is readily handled in calm or light winds, its bulk makes it difficult to control in heavy winds.

The Element of Danger.—Numerous balloons have been destroyed by lightning and similar causes. One of the largest of the Zeppelins was thus lost at Stuttgart in 1908.

Some Balloon Performances.

It is only a matter of fairness to state that, under favorable conditions, some very creditable records have been made with modern balloons, viz:

November 23d, 1907, the French dirigible Patrie, travelled 187 miles in 6 hours and 45 minutes against a light wind. This was a little over 28 miles an hour.

The Clement-Bayard, another French machine, sold to the Russian government, made a trip of 125 miles at a rate of 27 miles an hour.

Zeppelin No. 3, carrying eight passengers, and having a total lifting capacity of 5,500 pounds of ballast in addition to passengers, weight of equipment, etc., was tested in October, 1906, and made 67 miles in 2 hours and 17 minutes, about 30 miles an hour.

These are the best balloon trips on record, and show forcefully the limitations of speed, the greatest being not over 30 miles an hour.

Speed of Flying Machines.

Opposed to the balloon performances we have flying machine trips (of authentic records) as follows:

Bleriot—monoplane—in 1908—52 miles an hour.

Delagrange—June 22, 1908—10 1/2 miles in 16 minutes, approximately 42 miles an hour.

Wrights—October, 1905—the machine was then in its infancy—24 miles in 38 minutes, approximately 44 miles an hour. On December 31, 1908, the Wrights made 77 miles in 2 hours and 20 minutes.

Lambert, a pupil of the Wrights, and using a Wright biplane, on October 18, 1909, covered 29.82 miles in 49 minutes and 39 seconds, being at the rate of 36 miles an hour. This flight was made at a height of 1,312 feet.

Latham—October 21, 1909—made a short flight, about 11 minutes, in the teeth of a 40 mile gale, at Blackpool, Eng. He used an Antoniette monoplane, and the official report says: "This exhibition of nerve, daring and ability is unparalled in the history of aviation."

Farman—October 20, 1909—was in the air for 1 hour, 32 min., 16 seconds, travelling 47 miles, 1,184 yards, a duration record for England.

Paulhan—January 18, 1901—47 1/2 miles at the rate of 45 miles an hour, maintaining an altitude of from 1,000 to 2,000 feet.

Expense of Producing Gas.

Gas is indispensable in the operation of dirigible balloons, and gas is expensive. Besides this it is not always possible to obtain it in sufficient quantities even in large cities, as the supply on hand is generally needed for regular customers. Such as can be had is either water or coal gas, neither of which is as efficient in lifting power as hydrogen.

Hydrogen is the lightest and consequently the most buoyant of all known gases. It is secured commercially by treating zinc or iron with dilute sulphuric or hydrochloric acid. The average cost may be safely placed at $10 per 1,000 feet so that, to inflate a balloon of the size of the Zeppelin, holding 460,000 cubic feet, would cost $4,600.

Proportions of Materials Required.

In making hydrogen gas it is customary to allow 20 per cent for loss between the generation and the introduction of the gas into the balloon. Thus, while the formula calls for iron 28 times heavier than the weight of the hydrogen required, and acid 49 times heavier, the real quantities are 20 per cent greater. Hydrogen weighs about 0.09 ounce to the cubic foot. Consequently if we need say 450,000 cubic feet of gas we must have 2,531.25 pounds in weight. To produce this, allowing for the 20 percent loss, we must have 35 times its weight in iron, or over 44 tons. Of acid it would take 60 times the weight of the gas, or nearly 76 tons.

In Time of Emergency.

These figures are appalling, and under ordinary conditions would be prohibitive, but there are times when the balloon operator, unable to obtain water or coal gas, must foot the bills. In military maneuvers, where the field of operation is fixed, it is possible to furnish supplies of hydrogen gas in portable cylinders, but on long trips where sudden leakage or other cause makes descent in an unexpected spot unavoidable, it becomes a question of making your own hydrogen gas or deserting the balloon. And when this occurs the balloonist is up against another serious proposition—can he find the necessary zinc or iron? Can he get the acid?

Balloons for Commercial Use.

Despite all this the balloon has its uses. If there is to be such a thing as aerial navigation in a commercial way—the carrying of freight and passengers—it will come through the employment of such monster balloons as Count Zeppelin is building. But even then the carrying capacity must of necessity be limited. The latest Zeppelin creation, a monster in size, is 450 feet long, and 42 1/2 feet in diameter. The dimensions are such as to make all other balloons look like pigmies; even many ocean-going steamers are much smaller, and yet its passenger capacity is very small. On its 36-hour flight in May, 1909, the Zeppelin, carried only eight passengers. The speed, however, was quite respectable, 850 miles being covered in the 36 hours, a trifle over 23 miles an hour. The reserve buoyancy, that is the total lifting capacity aside from the weight of the airship and its equipment, is estimated at three tons.

CHAPTER XXII. PROBLEMS OF AERIAL FLIGHT.

In a lecture before the Royal Society of Arts, reported in Engineering, F. W. Lanchester took the position that practical flight was not the abstract question which some apparently considered it to be, but a problem in locomotive engineering. The flying machine was a locomotive appliance, designed not merely to lift a weight, but to transport it elsewhere, a fact which should be sufficiently obvious. Nevertheless one of the leading scientific men of the day advocated a type in which this, the main function of the flying machine, was overlooked. When the machine was considered as a method of transport, the vertical screw type, or helicopter, became at once ridiculous. It had, nevertheless, many advocates who had some vague and ill-defined notion of subsequent motion through the air after the weight was raised.

Helicopter Type Useless.

When efficiency of transport was demanded, the helicopter type was entirely out of court. Almost all of its advocates neglected the effect of the motion of the machine through the air on the efficiency of the vertical screws. They either assumed that the motion was so slow as not to matter, or that a patch of still air accompanied the machine in its flight. Only one form of this type had any possibility of success. In this there were two screws running on inclined axles—one on each side of the weight to be lifted. The action of such inclined screw was curious, and in a previous lecture he had pointed out that it was almost exactly the same as that of a bird's wing. In high-speed racing craft such inclined screws were of necessity often used, but it was at a sacrifice of their efficiency. In any case the efficiency of the inclined-screw helicopter could not compare with that of an aeroplane, and that type might be dismissed from consideration so soon as efficiency became the ruling factor of the design.

Must Compete With Locomotive.

To justify itself the aeroplane must compete, in some regard or other, with other locomotive appliances, performing one or more of the purposes of locomotion more efficiently than existing systems. It would be no use unless able to stem air currents, so that its velocity must be greater than that of the worst winds liable to be encountered. To illustrate the limitations imposed on the motion of an aeroplane by wind velocity, Mr. Lanchester gave the diagrams shown in Figs. 1 to 4. The circle in each case was, he said, described with a radius equal to the speed of the aeroplane in still air, from a center placed "down-wind" from the aeroplane by an amount equal to the velocity of the wind.

Fig. 1 therefore represented the case in which the air was still, and in this case the aeroplane represented by A had perfect liberty of movement in any direction

In Fig. 2 the velocity of the wind was half that of the aeroplane, and the latter could still navigate in any direction, but its speed against the wind was only one-third of its speed with the wind.

In Fig. 3 the velocity of the wind was equal to that of the aeroplane, and then motion against the wind was impossible; but it could move to any point of the circle, but not to any point lying to the left of the tangent AB. Finally, when the wind had a greater speed than the aeroplane, as in Fig. 4, the machine could move only in directions limited by the tangents AC and AD.

Matter of Fuel Consumption.

Taking the case in which the wind had a speed equal to half that of the aeroplane, Mr. Lanchester said that for a given journey out and home, down wind and back, the aeroplane would require 30 per cent more fuel than if the trip were made in still air; while if the journey was made at right angles to the direction of the wind the fuel needed would be 15 per cent more than in a calm. This 30 per cent extra was quite a heavy enough addition to the fuel; and to secure even this figure it was necessary that the aeroplane should have a speed of twice that of the maximum wind in which it was desired to operate the machine. Again, as stated in the last lecture, to insure the automatic stability of the machine it was necessary that the aeroplane speed should be largely in excess of that of the gusts of wind liable to be encountered.

Eccentricities of the Wind.

There was, Mr. Lanchester said, a loose connection between the average velocity of the wind and the maximum speed of the gusts. When the average speed of the wind was 40 miles per hour, that of the gusts might be equal or more. At one moment there might be a calm or the direction of the wind even reversed, followed, the next moment, by a violent gust. About the same minimum speed was desirable for security against gusts as was demanded by other considerations. Sixty miles an hour was the least figure desirable in an aeroplane, and this should be exceeded as much as possible. Actually, the Wright machine had a speed of 38 miles per hour, while Farman's Voisin machine flew at 45 miles per hour.

Both machines were extremely sensitive to high winds, and the speaker, in spite of newspaper reports to the contrary, had never seen either flown in more than a gentle breeze. The damping out of the oscillations of the flight path, discussed in the last lecture, increased with the fourth power of the natural velocity of flight, and rapid damping formed the easiest, and sometimes the only, defense against dangerous oscillations. A machine just stable at 35 miles per hour would have reasonably rapid damping if its speed were increased to 60 miles per hour.

Thinks Use Is Limited.

It was, the lecturer proceeded, inconceivable that any very extended use should be made of the aeroplane unless the speed was much greater than that of the motor car. It might in special cases be of service, apart from this increase of speed, as in the exploration of countries destitute of roads, but it would have no general utility. With an automobile averaging 25 to 35 miles per hour, almost any part of Europe, Russia excepted, was attainable in a day's journey. A flying machine of but equal speed would have no advantages, but if the speed could be raised to 90 or 100 miles per hour, the whole continent of Europe would become a playground, every part being within a daylight flight of Berlin. Further, some marine craft now had speeds of 40 miles per hour, and efficiently to follow up and report movements of such vessels an aeroplane should travel at 60 miles per hour at least. Hence from all points of view appeared the imperative desirability of very high velocities of flight. The difficulties of achievement were, however, great.

Weight of Lightest Motors.

As shown in the first lecture of his course, the resistance to motion was nearly independent of the velocity, so that the total work done in transporting a given weight was nearly constant. Hence the question of fuel economy was not a bar to high velocities of flight, though should these become excessive, the body resistance might constitute a large proportion of the total. The horsepower required varied as the velocity, so the factor governing the maximum velocity of flight was the horsepower that could be developed on a given weight. At present the weight per horsepower of feather-weight motors appeared to range from 2 1/4 pounds up to 7 pounds per brake horsepower, some actual figures being as follows:

> Antoinette......... 5 lbs. Fiat.............. 3 lbs. Gnome....... Under 3 lbs. Metallurgic....... 8 lbs. Renault........... 7 lbs. Wright.............6 lbs.

Automobile engines, on the other hand, commonly weighed 12 pounds to 13 pounds per brake horsepower.

For short flights fuel economy was of less importance than a saving in the weight of the engine. For long flights, however, the case was different. Thus, if the gasolene consumption was 1/2 pound per horsepower hour, and the engine weighed 3 pounds per brake horsepower, the fuel needed for a six-hour flight would weigh as much as the engine, but for half an hour's flight its weight would be unimportant.

Best Means of Propulsion.

The best method of propulsion was by the screw, which acting in air was subject to much the same conditions as obtained in marine work. Its efficiency depended on its diameter and pitch and on its position, whether in front of or behind the body propelled. From this theory of dynamic support, Mr. Lanchester proceeded, the efficiency of each element of a screw propeller could be represented by curves such as were given in his first lecture before the society, and from these curves the over-all efficiency of any proposed propeller could be computed, by mere inspection, with a fair degree of accuracy. These curves showed that the tips of long-bladed propellers were inefficient, as was also the portion of the blade near the root. In actual marine practice the blade from boss to tip was commonly of such a length that the over-all efficiency was 95 per cent of that of the most efficient element of it.

Advocates Propellers in Rear.

From these curves the diameter and appropriate pitch of a screw could be calculated, and the number of revolutions was then fixed. Thus, for a speed of 80 feet per second the pitch might come out as 8 feet, in which case the revolutions would be 600 per minute, which might, however, be too low for the motor. It was then necessary either to gear down the propeller, as was done in the Wright machine, or, if it was decided to drive it direct, to sacrifice some of the efficiency of the propeller. An analogous case arose in the application of the steam turbine to the propulsion of cargo boats, a problem as yet unsolved. The propeller should always be aft, so that it could abstract energy from the wake current, and also so that its wash was clear of the body propelled. The best possible efficiency was about 70 per cent, and it was safe to rely upon 66 per cent.

Benefits of Soaring Flight.

There was, Mr. Lanchester proceeded, some possibility of the aeronaut reducing the power needed for transport by his adopting the principle of soaring flight, as exemplified by some birds. There were, he continued, two different modes of soaring flight. In the one the bird made use of the upward current of air often to be found in the neighborhood of steep vertical cliffs. These cliffs deflected the air upward long before it actually reached the cliff, a whole region below being thus the seat of an upward current. Darwin has noted that the condor was only to be found in the neighborhood of such cliffs. Along the south coast also the gulls made frequent use of the up currents due to the nearly perpendicular chalk cliffs along the shore.

In the tropics up currents were also caused by temperature differences. Cumulus clouds, moreover, were nearly always the terminations of such up currents of heated air, which, on cooling by expansion in the upper regions, deposited their moisture as fog. These clouds might, perhaps, prove useful in the future in showing the aeronaut where up currents were to be found. Another mode of soaring flight was that adopted by the albatross, which took advantage of the fact that the air moved in pulsations, into which the bird fitted itself, being thus able to extract energy from the wind.

Whether it would be possible for the aeronaut to employ a similar method must be left to the future to decide.

Main Difficulties in Aviation.

In practical flight difficulties arose in starting and in alighting. There was a lower limit to the speed at which the machine was stable, and it was inadvisable to leave the ground till this limit was attained. Similarly, in alighting it was inexpedient to reduce the speed below the limit of stability. This fact constituted a difficulty in the adoption of high speeds, since the length of run needed increased in proportion to the square of the velocity. This drawback could, however, be surmounted by forming starting and alighting grounds of ample size. He thought it quite likely in the future that such grounds would be considered as essential to the flying machine as a seaport was to an ocean-going steamer or as a road was to the automobile.

Requisites of Flying Machine.

Flying machines were commonly divided into monoplanes and biplanes, according as they had one or two supporting surfaces. The distinction was not, however, fundamental. To get the requisite strength some form of girder framework was necessary, and it was a mere question of convenience whether the supporting surface was arranged along both the top and the bottom of this girder, or along the bottom only. The framework adopted universally was of wood braced by ties of pianoforte wire, an arrangement giving the stiffness desired with the least possible weight. Some kind of chassis was also necessary.

CHAPTER XXIII. AMATEURS MAY USE WRIGHT PATENTS.

Owing to the fact that the Wright brothers have enjoined a number of professional aviators from using their system of control, amateurs have been slow to adopt it. They recognize its merits, and would like to use the system, but have been apprehensive that it might involve them in litigation. There is no danger of this, as will be seen by the following statement made by the Wrights:

What Wright Brothers Say.

"Any amateur, any professional who is not exhibiting for money, is at liberty to use our patented devices. We shall be glad to have them do so, and there will be no interference on our part, by legal action, or otherwise. The only men we proceed against are those who, without our permission, without even asking our consent, coolly appropriate the results of our labors and use them for the purpose of making money. Curtiss, Delagrange, Voisin, and all the rest of them who have used our devices have done so in money-making exhibitions. So long as there is any money to be made by the use of the products of our brains, we propose to have it ourselves. It is the only way in which we can get any return for the years of patient work we have given to the problem of aviation. On the other hand, any man who wants to use these devices for the purpose of pleasure, or the advancement of science, is welcome to do so, without money and without price. This is fair enough, is it not?"

Basis of the Wright Patents.

In a flying machine a normally flat aeroplane having lateral marginal portions capable of movement to different positions above or below the normal plane of the body of the aeroplane, such movement being about an axis transverse to the line of flight, whereby said lateral marginal portions may be moved to different angles relatively to the normal plane of the body of the aeroplane, so as to present to the atmosphere different angles of incidence, and means for so moving said lateral marginal portions, substantially as described.

Application of vertical struts near the ends having flexible joints.

Means for simultaneously imparting such movement to said lateral portions to different angles relatively to each other.

Refers to the movement of the lateral portions on the same side to the same angle.

Means for simultaneously moving vertical rudder so as to present to the wind that side thereof nearest the side of the aeroplane having the smallest angle of incidence.

Lateral stability is obtained by warping the end wings by moving the lever at the right hand of the operator, connection being made by wires from the lever to the wing tips. The rudder may also be curved or warped in similar manner by lever action.

Wrights Obtain an Injunction.

In January, 1910, Judge Hazel, of the United States Circuit Court, granted a preliminary injunction restraining the Herring-Curtiss Co., and Glenn H. Curtiss, from manufacturing, selling, or using for exhibition purposes the machine known as the Curtiss aeroplane. The injunction was obtained on the ground that the Curtiss machine is an infringement upon the Wright patents in the matter of wing warping and rudder control.

It is not the purpose of the authors to discuss the subject pro or con. Such discussion would have no proper place in a volume of this kind. It is enough to say that Curtiss stoutly insists that his machine is not an infringement of the Wright patents, although Judge Hazel evidently thinks differently.

What the Judge Said.

In granting the preliminary injunction the judge said:

"Defendants claim generally that the difference in construction of their apparatus causes the equilibrium or lateral balance to be maintained and its aerial movement secured upon an entirely different principle from that of complainant; the defendants' aeroplanes are curved, firmly attached to the stanchions and hence are incapable of twisting or turning in any direction; that the supplementary planes or so-called rudders are secured to the forward stanchion at the extreme lateral ends of the planes and are adjusted midway between the upper and lower planes with the margins extending beyond the edges; that in moving the supplementary planes equal and uniform angles of incidence are presented as distinguished from fluctuating angles of incidence. Such claimed functional effects, however, are strongly contradicted by the expert witness for complainant.

Similar to Plan of Wrights.

"Upon this contention it is sufficient to say that the affidavits for the complainant so clearly define the principle of operation of the flying machines in question that I am reasonably satisfied that there is a variableness of the angle of incidence in the machine of defendants which is produced when a supplementary plane on one side is tilted or raised and the other stimultaneously tilted or lowered. I am also satisfied that the rear rudder is turned by the operator to the side having the least angle of incidence and that such turning is done at the time the supplementary planes are raised or depressed to prevent tilting or upsetting the machine. On the papers presented I incline to the view, as already indicated, that the claims of the patent in suit should be broadly construed; and when given such construction, the elements of the Wright machine are found in defendants' machine performing the same functional result. There are dissimilarities in the defendants' structure—changes of form and strengthening of parts—which may be improvements, but such dissimilarities seem to me to have no bearing upon the means adopted to preserve the equilibrium, which means are the equivalent of the claims in suit and attain an identical result.

Variance From Patent Immaterial.

"Defendants further contend that the curved or arched surfaces of the Wright aeroplanes in commercial use are departures from the patent, which describes 'substantially flat surfaces,' and that such a construction would be wholly impracticable. The drawing, Fig. 3, however, attached to the specification, shows a curved line inward of the aeroplane with straight lateral edges, and considering such drawing with the terminology of the specification, the slight arching of the surface is not thought a material departure; at any rate, the patent in issue does not belong to the class of patents which requires narrowing to the details of construction."

"June Bug" First Infringement.

Referring to the matter of priority, the judge said:

"Indeed, no one interfered with the rights of the patentees by constructing machines similar to theirs until in July, 1908, when Curtiss exhibited a flying machine which he called the 'June Bug.' He was immediately notified by the patentees that such machine with its movable surfaces at the tips of wings infringed the patent in suit, and he replied that he did not intend to publicly exhibit the machine for profit, but merely was engaged in exhibiting it for scientific purposes as a member of the Aerial Experiment Association. To this the patentees did not object. Subsequently, however, the machine, with supplementary planes placed midway between the upper and lower aeroplanes, was publicly exhibited by the defendant corporation and used by Curtiss in aerial flights for prizes and emoluments. It further appears that the defendants now threaten to continue such use for gain and profit, and to engage in the manufacture and sale of such infringing machines, thereby becoming an active rival of complainant in the business of constructing flying machines embodying the claims in suit, but such use of the infringing machines it is the duty of this court, on the papers presented, to enjoin.

"The requirements in patent causes for the issuance of an injunction pendente lite—the validity of the patent, general acquiescence by the public and infringement by the defendants—are so reasonably clear that I believe if not probable the complainant may succeed at final hearing, and therefore, status quo should be preserved and a preliminary injunction granted.

"So ordered."

Points Claimed By Curtiss.

That the Herring-Curtiss Co. will appeal is a certainty. Mr. Emerson R. Newell, counsel for the company, states its case as follows:

"The Curtiss machine has two main supporting surfaces, both of which are curved * * * and are absolutely rigid at all times and cannot be moved, warped or distorted in any manner. The front horizontal rudder is used for the steering up or down, and the rear vertical rudder is used only for steering to the right or left, in the same manner as a boat is steered by its rudder. The machine is provided at the rear with a fixed horizontal surface, which is not present in the machine of the patent, and which has a distinct advantage in the operation of defendants' machine, as will be hereafter discussed.

Does Not Warp Main Surface.

"Defendants' machine does not use the warping of the main supporting surfaces in restoring the lateral equilibrium, but has two comparatively small pivoted balancing surfaces or rudders. When one end of the machine is tipped up or down from the normal, these planes may be thrown in opposite directions by the operator, and so steer each end of the machine up or down to its normal level, at which time tension upon them is released and they are moved back by the pressure of the wind to their normal position.

Rudder Used Only For Steering.

"When defendants' balancing surfaces are moved they present equal angles of incidence to the normal rush of air and equal resistances, at each side of the machine, and there is therefore no tendency to turn around a vertical axis as is the case of the machine of the patent, consequently no reason or necessity for turning the vertical rear rudder in defendants' machine to counteract any such turning tendency. At any rate, whatever may be the theories in regard to this matter, the fact is that the operator of defendants' machine does not at any time turn his vertical rudder to counteract any turning tendency clue to the side balancing surfaces, but only uses it to steer the machine the same as a boat is steered."

Aero Club Recognizes Wrights.

The Aero Club of America has officially recognized the Wright patents. This course was taken following a conference held April 9th, 1910, participated in by William Wright and Andrew Freedman, representing the Wright Co., and the Aero Club's committee, of Philip T. Dodge, W. W. Miller, L. L. Gillespie, Wm. H. Page and Cortlandt F. Bishop.

At this meeting arrangements were made by which the Aero Club recognizes the Wright patents and will not give its section to any open meet where the promoters thereof have not secured a license from the Wright Company.

The substance of the agreement was that the Aero Club of America recognizes the rights of the owners of the Wright patents under the decisions of the Federal courts and refuses to countenance the infringement of those patents as long as these decisions remain in force.

In the meantime, in order to encourage aviation, both at home and abroad, and in order to permit foreign aviators to take part in aviation contests in this country it was agreed that the Aero Club of America, as the American representative of the International Aeronautic Federation, should approve only such public contests as may be licensed by the Wright Company and that the Wright Company, on the other hand, should encourage the holding of open meets or contests where ever approved as aforesaid by the Aero Club of America by granting licenses to promoters who make satisfactory arrangements with the company for its compensation for the use of its patents. At such licensed meet any machine of any make may participate freely without securing any further license or permit. The details and terms of all meets will be arranged by the committee having in charge the interests of both organizations.

CHAPTER XXIV. HINTS ON PROPELLER CONSTRUCTION.

Every professional aviator has his own ideas as to the design of the propeller, one of the most important features of flying-machine construction. While in many instances the propeller, at a casual glance, may appear to be identical, close inspection will develop the fact that in nearly every case some individual idea of the designer has been incorporated. Thus, two propellers of the two-bladed variety, while of the same general size as to length and width of blade, will vary greatly as to pitch and "twist" or curvature.

What the Designers Seek.

Every designer is seeking for the same result—the securing of the greatest possible thrust, or air displacement, with the least possible energy.

The angles of any screw propeller blade having a uniform or true pitch change gradually for every increased diameter. In order to give a reasonably clear explanation, it will be well to review in a primary way some of the definitions or terms used in connection with and applied to screw propellers.

Terms in General Use.

Pitch.—The term "pitch," as applied to a screw propeller, is the theoretical distance through which it would travel without slip in one revolution, and as applied to a propeller blade it is the angle at which the blades are set so as to enable them to travel in a spiral path through a fixed distance theoretically without slip in one revolution.

Pitch speed.—The term "pitch speed" of a screw propeller is the speed in feet multiplied by the number of revolutions it is caused to make in one minute of time. If a screw propeller is revolved 600 times per minute, and if its pitch is 7 ft., then the pitch speed of such a propeller would be 7x600 revolutions, or 4200 ft. per minute.

Uniform pitch.—A true pitch screw propeller is one having its blades formed in such a manner as to enable all of its useful portions, from the portion nearest the hub to its outer portion, to travel at a uniform pitch speed. Or, in other words, the pitch is uniform when the projected area of the blade is parallel along its full length and at the same time representing a true sector of a circle.

All screw propellers having a pitch equal to their diameters have the same angle for their blades at their largest diameter.

When Pitch Is Not Uniform.

A screw propeller not having a uniform pitch, but having the same angle for all portions of its blades, or some arbitrary angle not a true pitch, is distinguished from one having a true pitch in the variation of the pitch speeds that the various portions of its blades are forced to travel through while traveling at its maximum pitch speed.

On this subject Mr. R. W. Jamieson says in Aeronautics:

"Take for example an 8-foot screw propeller having an 8-foot pitch at its largest diameter. If the angle is the same throughout its entire blade length, then all the porions of its blades approaching the hub from its outer portion would have a gradually decreasing pitch. The 2-foot portion would have a 2-foot pitch; the 3-foot portion a 3-foot pitch, and so on to the 8-foot portion which would have an 8-foot pitch. When this form of propeller is caused to revolve, say 500 r.p.m., the 8-foot portion would have a calculated pitch speed of 8 feet by 500 revolutions, or 4,000 feet per min.; while the 2-foot portion would have a calculated pitch speed of 500 revolutions by 2 feet, or 1,000 feet per minute.

Effect of Non-Uniformity.

"Now, as all of the portions of this type of screw propeller must travel at some pitch speed, which must have for its maximum a pitch speed in feet below the calculated pitch speed of the largest diameter, it follows that some portions of its blades would perform useful work while the action of the other portions would be negative—resisting the forward motion of the portions having a greater pitch speed. The portions having a pitch speed below that at which the screw is traveling cease to perform useful work after their pitch speed has been exceeded by the portions having a larger diameter and a greater pitch speed.

"We might compare the larger and smaller diameter portions of this form of screw propeller, to two power-driven vessels connected with a line, one capable of traveling 20 miles per hour, the other 10 miles per hour. It can be readily understood that the boat capable of traveling 10 miles per hour would have no useful effect to help the one traveling 20 miles per hour, as its action would be such as to impose a dead load upon the latter's progress."

The term "slip," as applied to a screw propeller, is the distance between its calculated pitch speed and the actual distance it travels through under load, depending upon the efficiency and proportion of its blades and the amount of load it has to carry.

The action of a screw propeller while performing useful work might be compared to a nut traveling on a threaded bolt; little resistance is offered to its forward motion while it spins freely without load, but give it a load to carry; then it will take more power to keep up its speed; if too great a load is applied the thread will strip, and so it is with a screw propeller gliding spirally on the air. A propeller traveling without load on to new air might be compared to the nut traveling freely on the bolt. It would consume but little power and it would travel at nearly its calculated pitch speed, but give it work to do and then it will take power to drive it.

There is a reaction caused from the propeller projecting air backward when it slips, which, together with the supporting effect of the blades, combine to produce useful work or pull on the object to be carried.

A screw propeller working under load approaches more closely to its maximum efficiency as it carries its load with a minimum amount of slip, or nearing its calculated pitch speed.

Why Blades Are Curved.

It has been pointed out by experiment that certain forms of curved surfaces as applied to aeroplanes will lift more per horse power, per unit of square foot, while on the other hand it has been shown that a flat surface will lift more per horse power, but requires more area of surface to do it.

As a true pitch screw propeller is virtually a rotating aeroplane, a curved surface may be advantageously employed when the limit of size prevents using large plane surfaces for the blades.

Care should be exercised in keeping the chord of any curve to be used for the blades at the proper pitch angle, and in all cases propeller blades should be made rigid so as to preserve the true angle and not be distorted by centrifugal force or from any other cause, as flexibility will seriously affect their pitch speed and otherwise affect their efficiency.

How to Determine Angle.

To find the angle for the proper pitch at any point in the diameter of a propeller, determine the circumference by multiplying the diameter by 3.1416, which represent by drawing a line to scale in feet. At the end of this line draw another line to represent the desired pitch in feet. Then draw a line from the point representing the desired pitch in feet to the beginning of the circumference line. For example:

If the propeller to be laid out is 7 feet in diameter, and is to have a 7-foot pitch, the circumference will be 21.99 feet. Draw a diagram representing the circumference line and pitch in feet. If this diagram is wrapped around a cylinder the angle line will represent a true thread 7 feet in diameter and 7 feet long, and the angle of the thread will be 17 3/4 degrees.

Relation of Diameter to Circumference.

Since the areas of circles decrease as the diameter lessens, it follows that if a propeller is to travel at a uniform pitch speed, the volume of its blade displacement should decrease as its diameter becomes less, so as to occupy a corresponding relation to the circumferences of larger diameters, and at the same time the projected area of the blade must be parallel along its full length and should represent a true sector of a circle.

Let us suppose a 7-foot circle to be divided into 20 sectors, one of which represents a propeller blade. If the pitch is to be 7 feet, then the greatest depth of the angle would be 1/20 part of the pitch, or 4 2/10 inch. If the line representing the greatest depth of the angle is kept the same width as it approaches the hub, the pitch will be uniform. If the blade is set at an angle so its projected area is 1/20 part of the pitch, and if it is moved through 20 divisions for one revolution, it would have a travel of 7 feet.

CHAPTER XXV. NEW MOTORS AND DEVICES.

Since the first edition of this book was printed, early in 1910, there has been a remarkable advance in the construction of aeroplane motors, which has resulted in a wonderful decrease in the amount of surface area from that formerly required. Marked gain in lightness and speed of the motor has enabled aviators to get along, in some instances, with one-quarter of the plane supporting area previously used. The first Wright biplane, propelled by a motor of 25 h.p., productive of a fair average speed of 30 miles an hour, had a plane surface of 538 square feet. Now, by using a specially designed motor of 65 h. p., capable of developing a speed of from 70 to 80 miles an hour, the Wrights are enabled to successfully navigate a machine the plane area of which is about 130 square feet. This apparatus is intended to carry only one person (the operator). At Belmont Park, N. Y., the Wrights demonstrated that the small-surfaced biplane is much faster, easier to manage in the hands of a skilled manipulator, and a better altitude climber than the large and cumbersome machines with 538 square feet of surface heretofore used by them.

In this may be found a practical illustration of the principle that increased speed permits of a reduction in plane area in mathematical ratio to the gain in speed. The faster any object can be made to move through the air, the less will be the supporting surface required to sustain a given weight. But, there is a limit beyond which the plane surface cannot be reduced with safety. Regard must always be had to the securing of an ample sustaining surface so that in case of motor stoppage there will be sufficient buoyancy to enable the operator to descend safely.

The baby Wright used at the Belmont Park (N. Y.) aviation meet in the fall of 1910, had a plane length of 19 feet 6 inches, and an extreme breadth of 21 feet 6 inches, with a total surface area of 146 square feet. It was equipped with a new Wright 8-cylinder motor of 60 h. p., and two Wright propellers of 8 feet 6 inches diameter and 500 r. p. m. It was easily the fastest machine at the meet. After the tests, Wilbur Wright said:

"It is our intention to put together a machine with specially designed propellers, specially designed gears and a motor which will give us 65 horsepower at least. We will then be able, after some experimental work we are doing now, to send forth a machine that will make a new speed record."

In the new Wright machines the front elevating planes for up-and-down control have been eliminated, and the movements of the apparatus are now regulated solely by the rear, or "tail" control.

A Powerful Light Motor.

Another successful American aviation motor is the aeromotor, manufactured by the Detroit Aeronautic Construction. Aeromotors are made in four models as follows:

Model 1.—4-cylinder, 30-40 h. p., weight 200 pounds.

Model 2.—4-cylinder, (larger stroke and bore) 40-50 h. p., weight 225 pounds.

Model 3.—6-cylinder. 50-60 h. p., weight 210 pounds.

Model 4.—6-cylinder, 60-75 h. p., weight 275 pounds.

This motor is of the 4-cycle, vertical, water-cooled type. Roberts Aviation Motor.

One of the successful aviation motors of American make, is that produced by the Roberts Motor Co., of Sandusky, Ohio. It is designed by E. W. Roberts, M. E., who was formerly chief assistant and designer for Sir Hiram Maxim, when the latter was making his celebrated aeronautical experiments in England in 1894-95. This motor is made in both the 4- and 6-cylinder forms. The 4-cylinder motor weighs complete with Bosch magneto and carbureter 165 pounds, and will develop 40 actual brake h. p. at 1,000 r. p. m., 46 h. p. at 1,200 and 52 h. p. at 1,400. The 6-cylinder weighs 220 pounds and will develop 60 actual brake h. p. at 1,000 r. p. m., 69 h. p. at 1,200 and 78 h. p. at 1,500.

Extreme lightness has been secured by doing away with all superfluous parts, rather than by a shaving down of materials to a dangerous thinness. For example, there is neither an intake or exhaust manifold on the motor. The distributing valve forms a part of the crankcase as does the water intake, and the gear pump. Magnalium takes the place of aluminum in the crankcase, because it is not only lighter but stronger and can be cast very thin. The crankshaft is 2

1/2-inch diameter with a 2 1/4-inch hole, and while it would be strong enough in ordinary 40 per cent carbon steel it is made of steel twice the strength of that customarily employed. Similar care has been exercised on other parts and the result is a motor weighing 4 pounds per h. p.

The Rinek Motor.

The Rinek aviation motor, constructed by the Rinek Aero Mfg. Co., of Easton, Pa., is another that is meeting with favor among aviators. Type B-8 is an 8-cylinder motor, the cylinders being set at right angles, on a V-shaped crank case. It is water cooled, develops 50-60 h. p., the minimum at 1,220 r. p. m., and weighs 280 pounds with all accessories. Type B-4, a 4-cylinder motor, develops 30 h. p. at 1,800 r. p. m., and weighs 130 pounds complete. The cylinders in both motors are made of cast iron with copper water jackets.

The Overhead Camshaft Boulevard.

The overhead camshaft Boulevard is still another form of aviation motor which has been favorably received. This is the product of the Boulevard Engine Co., of St. Louis. It is made with 4 and 8 cylinders. The former develops 30-35 h. p. at 1,200 r. p. m., and weighs 130 pounds. The 8-cylinder motor gives 60-70 h. p. at 1,200 r. p. m., and weighs 200 pounds. Simplicity of construction is the main feature of this motor, especially in the manipulation of the valves.

CHAPTER XXVI. MONOPLANES, TRIPLANES, MULTIPLANES.

Until recently, American aviators had not given serious attention to any form of flying machines aside from biplanes. Of the twenty-one monoplanes competing at the International meet at Belmont Park, N. Y., in November, 1910, only three makes were handled by Americans. Moissant and Drexel navigated Bleriot machines, Harkness an Antoinette, and Glenn Curtiss a single decker of his own construction. On the other hand the various foreign aviators who took part in the meet unhesitatingly gave preference to monoplanes.

Whatever may have been the cause of this seeming prejudice against the monoplane on the part of American air sailors, it is slowly being overcome. When a man like Curtiss, who has attained great success with biplanes, gives serious attention to the monoplane form of construction and goes so far as to build and successfully operate a single surface machine, it may be taken for granted that the monoplane is a fixture in this country.

Dimensions of Monoplanes.

The makes, dimensions and equipment of the various monoplanes used at Belmont Park are as follows:

Bleriot—(Moissant, operator)—plane length 23 feet, extreme breadth 28 feet, surface area 160 square feet, 7-cylinder, 50 h. p. Gnome engine, Chauviere propeller, 7 feet 6 inches diameter, 1,200 r. p. m.

Bleriot—(Drexel, operator)—exactly the same as Moissant's machine.

Antoinette—(Harkness, operator)—plane length 42 feet, extreme breadth 46 feet, surface area 377 square feet, Emerson 6-cylinder, 50 h. p. motor, Antoinette propeller, 7 feet 6 inches diameter, 1,200 r. p. m.

Curtiss—(Glenn H. Curtiss, operator)—plane length 25 feet, extreme breadth 26 feet, surface area 130 square feet, Curtiss 8-cylinder, 60 h. p. motor, Paragon propeller, 7 feet in diameter, 1,200 r. p. m.

With one exception Curtiss had the smallest machine of any of those entering into competition. The smallest was La Demoiselle, made by Santos-Dumont, the proportions of which were: plane length 20 feet, extreme breadth 18 feet, surface area 100 square feet, Clement-Bayard 2-cylinder, 30 h. p. motor, Chauviere propeller, 6 feet 6 inches in diameter, 1,100 r. p. m.

Winnings Made with Monoplanes.

Operators of monoplanes won a fair share of the cash prizes. They won $30,283 out of a total of $63,250, to say nothing about Grahame-White's winnings. The latter won $13,600, but part of his winning flights were made in a Bleriot monoplane, and part in a Farman machine. Aside from Grahame-White the winnings were divided as follows: Moissant (Bleriot) $13,350; Latham (Antoinette) $8,183; Aubrun (Bleriot) $2,400; De Lesseps (Bleriot) $2,300; Drexel (Bleriot) $1,700; Radley (Bleriot) $1,300; Simon (Bleriot) $750; Andemars (Clement-Bayard) $100; Barrier (Bleriot) $100.

Out of a total of $30,283, operators of Bleriot machines won $21,900, again omitting Grahame-White's share. If the winnings with monoplane and biplane could be divided so as to show the amount won with each type of machine the credit side of the Bleriot account would be materially enlarged.

The Most Popular Monoplanes.

While the number of successful monoplanes is increasing rapidly, and there is some feature of advantage in nearly all the new makes, interest centers chiefly in the Santos-Dumont, Antoinette and Bleriot machines. This is because more has been accomplished with them than with any of the others, possibly because they have had greater opportunities.

For the guidance of those who may wish to build a machine of the monoplane type after the Santos-Dumont or Bleriot models, the following details will be found useful.

Santos-Dumont—The latest production of this maker is called the "No. 20 Baby." It is of 18 feet spread, and 20 feet over all in depth. It stands 4 feet 2 inches in height, not counting the propeller. When this latter is in a vertical position the extreme height of the machine is 7 feet 5 inches. It is strictly a one-man apparatus. The total surface area is 115 square feet. The total weight of the monoplane with engine and propeller is 352 pounds. Santos-Dumont weighs 110 pounds, so the entire weight carried while in flight is 462 pounds, or about 3.6 pounds per square foot of surface.

Bamboo is used in the construction of the body frame, and also for the frame of the tail. The body frame consists of three bamboo poles about 2 inches in diameter at the forward end and tapering to about 1 inch at the rear. These poles are jointed with brass sockets near the rear of the main plane so they may be taken apart easily for convenience in housing or transportation. The main plane is built upon four transverse spars of ash, set at a slight dihedral angle, two being placed on each side of the central bamboo. These spars are about 2 inches wide by 1 1/8-inch deep for a few feet each side of the center of the machine, and from there taper down to an inch in depth at the center bamboo, and at their outer ends, but the width remains the same throughout their entire length. The planes are double surfaced with silk and laced above and below the bamboo ribs which run fore and aft under the main spars and terminate in a forked clip through which a wire is strung for lacing on the silk. The tail consists of a horizontal and vertical surface placed on a universal joint about 10 feet back of the rear edge of the main plane. Both of these surfaces are flat and consist of a silk covering stretched upon bamboo ribs. The horizontal surface is 6 feet 5 inches across, and 4 feet 9 inches from front to back. The vertical surface is of the same width (6 feet 5 inches) but is only 3 feet 7 inches from front to back. All the details of construction are shown in the accompanying illustration.

Power is furnished by a very light (110 pounds) Darracq motor, of the double-opposed-cylinder type. It has a bore of 4.118 inches, and stroke of 4.724 inches, runs at 1,800 r. p. m., and with a 6 1/2-foot propeller develops a thrust of 242 1/2 pounds when the monoplane is held steady.

Bleriot—No. XI, the latest of the Bleriot productions, and the greatest record maker of the lot, is 28 feet in spread of main plane, and depth of 6 feet in largest part. This would give a main surface of 168 square feet, but as the ends of the plane are sharply tapered from the rear, the actual surface is reduced to 150 square feet. Projecting from the main frame is an elongated tail (shown in the illustration) which carries the horizontal and vertical rudders. The former is made in three sections. The center piece is 6 feet 1 inch in spread, and 2 feet 10 inches in depth, containing 17 square feet of surface. The end sections, which are made movable for warping purposes, are each 2 feet 10 inches square, the combined surface area in the entire horizontal rudder being 33 square feet. The vertical rudder contains 4 1/2 square feet of surface, making the entire supporting area 187 1/2 square feet.

From the outer end of the propeller shaft in front to the extreme rear edge of the vertical rudder, the machine is 25 feet deep. Deducting the 6-foot depth of the main plane leaves 19 feet as the length of the rudder beam and rudders. The motor equipment consists of a 3-cylinder, air-cooled engine of about 30 h. p. placed at the front end of the body frame, and carrying on its crankshaft a two-bladed propeller 6 feet 8 inches in diameter. The engine speed is about 1,250 r. p. m. at which the propeller develops a thrust of over 200 pounds.

The Bleriot XI complete weighs 484 pounds, and with operator and fuel supply ready for a 25- or 30-mile flight, 715 pounds. One peculiarity of the Bleriot construction is that, while the ribs of the main plane are curved, there is no preliminary bending of the pieces as in other forms of construction. Bleriot has his rib pieces cut a little longer than required and, by springing them into place, secures the necessary curvature. A good view of the Bleriot plane framework is given on page 63.

Combined Triplane and Biplane.

At Norwich, Conn., the Stebbins-Geynet Co., after several years of experiment, has begun the manufacture of a combination triplane and biplane machine. The center plane, which is located about midway between the upper and lower surfaces, is made removable. The change from triplane to biplane, or vice versa, may be readily made in a few minutes. The constructors claim for this type of air craft a large supporting surface area with the minimum of dimensions in planes. Although this machine has only 24-foot spread and is only 26 feet over all, its total amount of supporting area is 400 square feet; weight, 600 pounds in flying order, and lifting capacity approximately 700 pounds more.

The frame is made entirely of a selected grade of Oregon spruce, finished down to a smooth surface and varnished. All struts are fish-shaped and set in aluminum sockets, which are bolted to top and lower beams with special strong bolts of small diameter. The middle plane is set inside the six uprights and held in place by aluminum castings. A flexible twisted seven-strand wire cable and Stebbins-Geynet turnbuckles are used for trussing.

The top plane is in three sections, laced together. It has a 24-foot spread and is 7 feet in depth. The middle plane is in two sections each of 7 1/2 feet spread and 6 feet in depth. The center ends of the middle plane sections do not come within 5 feet of joining, this open space being left for the engine. The bottom plane is of 16 feet spread and 5 feet in depth. It will thus be seen that the planes overhang one another in depth, the bottom one being the smallest in this respect. The planes are set at an angle of 9 degrees, and there is a clear space of 3 1/2 feet between each, making the total distance from the bottom to the top plane a trifle over 7 feet. The total supporting surface in the main planes is 350 square feet. By arranging the three plane surfaces at an angle as described and varying their size, the greatest amount of lifting area is secured above the center of gravity, and the greatest weight carried below.

The ribs are made of laminated spruce, finished down to 1/2x3/4-inch cross section dimensions, with a curvature of about 1 in 20, and fastened to the beams with special aluminum castings. Number 2 Naiad aeroplane cloth is used in covering the planes, with pockets sewn in for the ribs.

Two combination elevating rudders are set up well in front, each having 18 square feet of supporting area. These rudders are arranged to work in unison, independently, or in opposite directions. In the Model B machine, there are also two small rear elevating rudders, which work in unison with the front rudders. One vertical rudder of 10 square feet is suspended in the rear of a small stationary horizontal plane in Model A, while the vertical rudder on Model B is only 6 square feet in size. The elevating rudders are arranged so as to act as stabilizing planes when the machine is in flight. The wing tips are held in place with a special two-piece casting which forms a hinge, and makes a quick detachable joint. Wing tips are also used in balancing.

Model A is equipped with a Cameron 25-30 h. p., 4-cylinder, air-cooled motor. On Model B a Holmes rotary 7-cylinder motor of 4x4-inch bore and stroke is used.

Positive control is secured by use of the Stebbins-Geynet "auto-control" system. A pull or push movement operates the elevating rudders, while the balancing is done by means of side movements or slight turns. The rear vertical rudder is manipulated by means of a foot lever.

New Cody Biplane.

Among the comparatively new biplanes is one constructed by Willard F. Cody, of London, Eng., the principal distinctive feature of which is an automatic control which works independently of the hand levers. For the other control a long lever carrying a steering wheel furnishes all the necessary control movements, there being no footwork at all. The lever is universally jointed and when moved fore and aft operates the two ailerons as if they were one; when the shaft is rotated it moves the tail as a whole. The horizontal tail component is immovable. When the lever is moved from side to side it works not only the ailerons and the independent elevators, but also through a peculiar arrangement, the vertical rear rudder as well.

The spread of the planes is 46 feet 6 inches and the width 6 feet 6 inches. The ailerons jut out 1 foot 6 inches on each side of the machine and are 13 feet 6 inches long. The cross-shaped tail is supported by an outrigger composed of two long bamboos and of this the vertical plane is 9 feet by 4 feet, while the horizontal plane is 8 feet by 4 feet. The over-all length of the machine is 36 feet. The lifting surface is 857 square feet. It will weigh, with a pilot, 1,450 pounds. The distance between the main planes is 8 feet 6 inches, which is a rather notable feature in this flyer.

The propeller has a diameter of 11 feet and 2 inches with a 13-foot 6-inch pitch; it is driven at 560 revolutions by a chain, and the gear reduction between the chain and propeller shaft is two to one.

The machine from elevator to tail plane bristles in original points. The hump in the ribs has been cut away entirely, so that although the plane is double surfaced, the surfaces are closest together at a point which approximates the center of pressure. The plane is practically of two stream-line forms, of which one is the continuation of the other. This construction, claims the inventor, will give increased lift, and decreased head resistance. The trials substantiate this, as the angle of incidence in flying is only about one in twenty-six.

The ribs in the main planes are made of strips of silver spruce one-half by one-half inch, while those in the ailerons are solid and one-fourth inch thick. In the main planes the fabric is held down with thin wooden fillets. Cody's planes are noted for their neatness, rigidity and smoothness. Pegamoid fabric is used throughout.

Pressey Automatic Control.

Another ingenious system of automatic control has been perfected by Dr. J. B. Pressey, of Newport News, Va. The aeroplane is equipped with a manually operated, vertical rudder, (3), at the stern, and a horizontal, manually operated, front control, (4), in front. At the ends of the main plane, and about midway between the upper and lower sections thereof, there are supplemental planes, (5).

In connection with these supplemental planes (5), there is employed a gravity influenced weight, the aviator in his seat, for holding them in a horizontal, or substantially horizontal, position when the main plane is traveling on an even keel; and for causing them to tip when the main plane dips laterally, to port or starboard, the planes (5) having a lifting effect upon the depressed end of the main plane, and a depressing effect upon the lifted end of the main plane, so as to correct such lateral dip of the main plane, and restore it to an even keel. To the forward, upper edge of planes (5) connection is made by means of rod (13) to one arm of a bellcrank lever, (14) the latter being pivotally mounted upon a fore and aft pin (15), supported from the main plane; and the other arms of the port and starboard bellcrank levers (16), are connected by rod (17), which has an eye (18), for receiving the segmental rod (19), secured to and projecting from cross bar on seat supporting yoke (7). When, therefore, the main plane tips downwardly on the starboard side, the rod

(17) will be moved bodily to starboard, and the starboard balancing plane (5) will be inclined so as to raise its forward edge and depress its rear edge, while, at the same time, the port balancing plane (5), will be inclined so as to depress its forward edge, and raise its rear edge, thereby causing the starboard balancing plane to exert a lifting effect, and the port balancing plane to exert a depressing effect upon the main plane, with the result of restoring the main plane to an even keel, at which time the balancing planes (5), will have resumed their normal, horizontal position.

When the main plane dips downwardly on the port side, a reverse action takes place, with the like result of restoring the main plane to an even keel. In order to correct forward and aft dip of the main plane, fore and aft balancing planes (20) and (23) are provided. These planes are carried by transverse rock shafts, which may be pivotally mounted in any suitable way, upon structures carried by main plane. In the present instance, the forward balancing plane is pivotally mounted in extensions (21) of the frame (22) which carries the forward, manually operated, horizontal ascending and descending plane

It is absolutely necessary, in making a turn with an aeroplane, if that turn is to be made in safety, that the main plane shall be inclined, or "banked," to a degree proportional to the radius of the curve and to the speed of the aeroplane. Each different curve, at the same speed, demands a different inclination, as is also demanded by each variation in speed in rounding like curves. This invention gives the desired result with absolute certainty.

The Sellers' Multiplane.

Another innovation is a multiplane, or four-surfaced machine, built and operated by M. B. Sellers, formerly of Grahn, Ky., but now located at Norwood, Ga. Aside from the use of four sustaining surfaces, the novelty in the Sellers machine lies in the fact that it is operated successfully with an 8 h. p. motor, which is the smallest yet used in actual flight. In describing his work, Mr. Sellers says his purpose has been to develop the efficiency of the surfaces to a point where flight may be obtained with the minimum of power and, judging by the results accomplished, he has succeeded. In a letter written to the authors of this book, Mr. Sellers says:

"I dislike having my machine called a quadruplane, because the number of planes is immaterial; the distinctive feature being the arrangement of the planes in steps; a better name would be step aeroplane, or step plane.

"The machine as patented, comprises two or more planes arranged in step form, the highest being in front. The machine I am now using has four planes 3 ft. x 18 ft.; total about 200 square feet; camber (arch) 1 in 16.

"The vertical keel is for lateral stability; the rudder for direction. This is the first machine (so far as I know) to have a combination of wheels and runners or skids (Oct. 1908). The wheels rise up automatically when the machine leaves the ground, so that it may alight on the runners.

"A Duthirt & Chalmers 2-cylinder opposed, 3 1/8-inch engine was used first, and several hundred short flights were made. The engine gave four brake h. p., which was barely sufficient for continued flight. The aeroplane complete with this engine weighed 78 pounds. The engine now used is a Bates 3 5/8-inch, 2-cylinder opposed, showing 8 h. p., and apparently giving plenty of power. The weight of aeroplane with this engine is now 110 pounds. Owing to poor grounds only short flights have been made, the longest to date (Dec. 31, 1910) being about 1,000 feet.

"In building the present machine, my object was to produce a safe, slow, light, and small h. p. aeroplane, a purpose which I have accomplished."

CHAPTER XXVII. 1911 AEROPLANE RECORDS.

THE WORLD AT LARGE.

Greatest Speed Per Hour, Whatever Length of Flight, Aviator Alone—E. Nieuport, Mourmelon, France, June 21, Nieuport Machine, 82.72 miles; with one passenger, E. Nieuport, Moumlelon, France, June 12, Nieuport Machine, 67.11 miles; with two passengers, E. Nieuport, Mourmelon, France, March 9, Nieuport Machine, 63.91 miles; with three passengers, G. Busson, Rheims, France, March 10, Deperdussin Machine, 59.84 miles; with four passengers, G. Busson, Rheims, France, March 10, Deperdussin Machine, 54.21 miles.

Greatest Distance Aviator Alone—G. Fourny, no stops, Buc, France, September 2, M. Farman Machine, 447.01 miles; E. Helen, three stops, Etampes, France, September 8, Nieuport Machine, 778.45 miles; with one passenger, Lieut. Bier, Austria, October 2, Etrich Machine, 155.34 miles; with two passengers, Lieut. Bier, Austria, October 4, Etrich Machine, 69.59 miles; with three passengers, G. Busson, Rheims, France, March 10, Deperdussin Machine, 31.06 miles; with four passengers, G. Busson, Rheims, France, March 10, Deperdussin Machine, 15.99 miles.

Greatest Duration Aviator Alone—G. Fourny, no stops, Buc, France, September 2, M. Farman Machine, 11 hours, 1 minute, 29 seconds, E. Helen, three stops, Etampes, France, September 8, Nieuport Machine, 14 hours, 7 minutes, 50 seconds, 13 hours, 17 minutes net time; with one passenger, Suvelack, Johannisthal, Germany, December 8, 4 hours, 23 minutes; with two passengers, T. de W. Milling, Nassau Boulevard, New York, September 26, Burgess-Wright Machine, 1 hour, 54 minutes, 42 3-5 seconds; with three passengers, Warchalowski, Wiener-Neustadt, Aust., October 30, 45 minutes, 46 seconds; with four passengers, G. Busson, Rheims, France, March 10, Deperdussin Machine, 17 minutes, 28 1-5 seconds.

Greatest Altitude Aviator Alone—Garros, St. Malo, France, September 4, Bleriot Machine, 13,362 feet; with one passenger, Prevost,

Courcy, France, December 2, 9,840 feet; with two passengers, Lieut. Bier, Austria, Etrich Machine, 4,010 feet.

AMERICAN RECORDS.

Greatest Speed Per Hour, Whatever Length of Flight, Aviator Alone—A. Leblanc, Belmont Park, N. Y., October 29, Bleriot Machine, 67.87 miles; with one passenger, C. Grahame-White, Squantum, Mass., September 4, Nieuport Machine, 63.23 miles; with two passengers, T. O. M. Sopwith, Chicago, Ill., August 15, Wright Machine, 34.96 miles.

Greatest Distance Aviator Alone—St. Croix Johnstone, Mineola, N. Y., July 27, Moisant (Bleriot Type) Machine, 176.23 miles.

Greatest Duration Aviator Alone—Howard W. Gill, Kinloch, Mo., October 19, Wright Machine, 4 hours, 16 minutes, 35 seconds; with one passenger, G. W. Beatty, Chicago, Ill., August 19, Wright Machine, 3 hours, 42 minutes, 22 1-5 seconds; with two passengers, T. de W. Milling, Nassau Boulevard, N. Y., September 26, Burgess-Wright Machine, 1 hour, 54 minutes, 42 3-5 seconds.

Greatest Altitude Aviator Alone—L. Beachy, Chicago, Ill., August 20, Curtiss Machine, 11,642 feet; with one passenger, C. Grahame-White, Nassau Boulevard, N. Y., September 30, Nieuport Machine, 3,347 feet.

Weight Carrying—P. O. Parmelee, Chicago, Ill., August 19, Wright Machine, 458 lbs.

AVIATION DEVELOPMENT.

The wonderful progress made in the science of aviation during the year 1911 far surpasses any twelve months' advancement recorded. The advancement has not been confined to any country or continent, since every part of the world is taking its part in aviation history making.

The rapidly increasing interest in aviation has brought forth schools for the instruction of flying in both the old and new world, and licensed air pilots before they receive their sanctions from the governing aero clubs of their country are required to pass an extremely trying examination in actual flights. Exhibition flights and races were common in all parts of the world during 1911, and touring aviators visited India, China, Japan, South Africa, Australia and South America, giving exhibitions and instruction.

Europe was the scene of a number of cross-country races in which entries ranging from ten to twenty aviators flew from city to city around a given circuit, which in some instances exceeded 1,000 miles in distance. Cross-country flights with and without passengers became so common that those of less than two hours' duration attracted little attention. There were fewer attempts at high altitude soaring, although the world's record in this department of aviation was bettered several times. In place of these high flights, the aviators devoted more attention to speed, duration and spectacular manoeuvres, which appeared to satisfy the spectators. The prize money won during 1911 exceeded $1,000,000, but owing to the increased number of aviators the individual winnings were not as large as in 1910.

It is estimated that within the past twelve months more than 300,000 miles have been covered in aeroplane flights and more than seven thousand persons, classed either as aviators or passengers, taken up into the air. The aeroplane of today ranges through monoplane, biplane, triplane and even quadraplane, and more than two hundred types of these machines are in use.

Aeroplanes are becoming a factor of international commerce. The records of the Bureau of Statistics show that more than $50,000 worth of aeroplanes were imported into, and exported from, the United States in the months of July, August and September, 1911. The Bureau of Statistics only began the maintenance of a separate record of this comparatively new article of commerce with the opening of the fiscal year 1911-12.

Two of the prominent developments of 1911 were the introduction of the hydro-aeroplane and the motorless glider experiments of the Wright brothers at Killdevil Hills, N. C., where during the two weeks' experiments numerous flights with and against the wind were made, culminating in the establishing of a record by Orville Wright on October 25, 1911, when in a 52-mile per hour blow he reached an elevation of 225 feet and remained in the air 10 minutes and 34 seconds. The search for the secret of automatic stability still continues, and though some remarkable progress has been made the solution has not yet been reached.

NOTABLE CROSS-COUNTRY FLIGHTS OF 1911.

One of the important features of 1911 in aviation was the rapid increase in the number and distance of cross-country flights made either for the purpose of exhibition, testing, instruction or pleasure. Flights between cities in almost every country of the world became common occurrences. So great was the number that only those of more than ordinary importance because of speed, distance or duration are recorded. The flights of Harry N. Atwood from Boston to Washington and from St. Louis to New York, and C. P. Rodgers from New York to Los Angeles were the most important events of the kind in this country. The St Louis to New York flight was a distance by air route, 1,266 miles. Duration of flight, 12 days. Net flying time, 28 hours 53 minutes. Average daily flight, 105.5 miles. Average speed, 43.9 miles per hour.

Transcontinental Flight of Calbraith P. Rodgers.—All world records for cross-country flying were broken during the New York to Los Angeles flight of Calbraith P. Rodgers, who left Sheepshead Bay, N. Y., on Sunday, September 17, 1911, and completed his flight to the Pacific Coast on Sunday, November 5, at Pasadena, Cal. Rodgers flew a Wright biplane, and during his long trip the machine was repeatedly repaired, so great was the strain of the long journey in the air. Rodgers is estimated to have covered 4,231 miles, although the actual route as mapped out was but 4,017 miles. Elapsed time to Pasadena, Cal., 49 days; actual time in the air, 4,924 minutes, equivalent to 3 days 10 hours 4 minutes; average speed approximating 51 miles per hour. Rodgers' longest flight in one day was from Sanderson to Sierra Blanca, Texas, on October 28, when he covered 231 miles. On November 12, Rodgers fell at Compton, Cal., and was badly injured, causing a delay of 28 days.

European Circuit Race.—Started from Paris on June 18, 1911. Distance, 1,073 miles, via Paris to Liege; Liege to Spa to Liege; Liege to Utrecht, Holland; Utrecht to Brussels, Belgium; Brussels to Roubaix; Roubaix to Calais; Calais to London; London to Calais and Calais to Paris. Three aeronauts were killed either at the start or shortly after the race was in progress. They were Capt. Princetau, M. Le Martin and M. Lendron. Three others were injured by falls. Seven hundred thousand spectators witnessed the start from the aviation field at

Vincennes, near Paris. There were more than forty starters, of which eight finished. The winner, Lieut. Jean Conneau, who flies under the name of "Andre Beaumont," completed the circuit on July 7; his actual net flying time for the distance being 58h. 38m. 4-5s.

Circuit of England Race—1,010 Miles in Five Sections.—

Start, July 22. Finish, July 26. Prize, $50,000. Twenty-eight entries and eighteen starters. Seventeen finished the first section from Brooklands to Hendon, a distance of twenty miles. Five reached Edinburgh, the second section, a distance of 343 miles, and four completed the entire circuit.

Paris to Madrid Race.—This race was started at the Paris aviation held at Issy-les-Moulineaux on Sunday, May 21. There were twenty-one entrants, and fully 300,000 spectators gathered to witness the initial flight of the aerial races. The race was divided into three stages as follows: Paris to Angouleme, 248 miles; Angouleme to St. Sebastian, 208 miles, and from St. Sebastian to Madrid, 386 miles, a total distance of 842 miles. After three of the entrants had safely left the field, Aviator Train lost control of his plane, and in falling struck and killed M. Berteaux, the French Minister of War, and seriously injured Premier Monis. The accident caused the withdrawal of all but six of the original entrants, and of these but one finished. The race called for a flight over the Pyrenees Mountains, and Vedrines, the winner, had to rise to a height of more than 7,000 feet to pass the mountain barrier near Somosierra Pass. Both Vedrines and Gibert, another competitor, were attacked by eagles during the latter stages of the flight. Vedrines, who started from Paris on Monday, May 22, finished the long and perilous race at 8:06 a. m. Friday, May 26. Vedrines net flying time, all controls and enforced stops subtracted, was 14h. 55m. 18s. The various prizes to the winner aggregated $30,000.

The Paris-Rome-Turin Race.—The conditions of this race called for a flight between the cities of Paris, Rome and Turin, covering a distance of 1,300 miles. The aviators were permitted by the rules to alight whenever and wherever they desired and the time limit was set from May 28 to June 15. A prize of $100,000 was offered the winner, but the contest was never finished, as one after another the aviators dropped out until Frey fell near Roncigilione, France, breaking both arms and legs and unofficially ending the contest. There were twenty-one entries and twelve actual starters.

International Speed Cup Race.—The third annual international James Gordon Bennett speed cup race was held at Eastchurch, England, on July 1, 1911, and for the second time was won by an American aviator, C. T. Weymann, in a French racing aeroplane. The distance was 150 kilometres equivalent to 94 miles, and the winner's time of 1h. 11m. 36s. showed an average speed of 78.77 miles per hour. The first race was held in 1909 and was won by Glenn Curtiss, who flew the twenty kilometres (12.4 miles) in 15 minutes 50 2-5 seconds at an average speed of 47 miles per hour. In 1910 the winner was Grahame-White, who covered 100 kilometres (62 miles) at Belmont Park, L. I., in 60 minutes 47 3-5 seconds, an average speed of 61.3 miles per hour. In the 1911 race there were six starters: three from France, two from Great Britain and one from the United States.

Milan to Turin to Milan Race.—This race which was started from Milan, Italy, on October 29, was restricted to Italian aviators and had six starters. The distance was approximately 177 miles and won by Manissero in a Bleriot machine in 3h. 16m. 2 4-5s.

New York to Philadelphia Race.—The first intercity aeroplane race ever held in the United States was started from New York City on August 5, and finished in Philadelphia the same day. The prize of $5,000 was offered by a commercial concern with stores in the two cities: Three entrants competed from the Curtiss Exhibition Company. The distance was approximately 83 miles and won by L. Beachey in a Curtiss machine in 1h. 50m. at an average speed of 45 miles per hour.

Tri-State Race.—The tri-state race was the feature event of the Harvard Aviation Society meet held at Squantum, Mass., August 26 to September 6. It was held Labor Day, September 4, over a course of 174 miles, from Boston to Nashua to Worcester to Providence to Boston. Four competitors started, of which two finished, the winner, E. Ovington, in a Bleriot machine. Ovington's net flying time, 3h. 6m. 22 1-5s. Winner's prize, $10,000.

AEROPLANES AND DIRIGIBLE BALLOONS IN WARFARE.

Wonderful progress has been made in the development of the aeroplane in this country and in Europe since 1903, and within the last two or three years the leading powers of the world have entered upon extensive tests and experiments to determine its availability and usefulness in land and naval warfare.

At the present time all the great powers are building or purchasing aeroplanes on an extensive scale. They have established government schools for the instruction of their army and navy officers and for experimental work. So-called "Airship Fleets" have been constructed and placed in commission as auxiliaries to the armies and navies. The fleets of France and Germany are about equal and are larger by far than those of any of the other powers. The length of the dirigibles composing these fleets runs from 150 to 500 feet; they are equipped with engines of from 50 to 500 horse-power, with a rate of speed ranging from 20 to 30 miles per hour. Their approximate range is from 200 to 900 miles; the longest actual run (made by the Zeppelin II, Germany) is 800 miles.

A British naval airship, one of the largest yet built, was completed last summer. It has cost over $200,000, and it was in course of designing and construction two years. It is 510 feet long; can carry 22 persons, and has a lift of 21 tons.

The relative value of the dirigible balloon and the aeroplane in actual war is yet to be determined. The dirigible is considered to be the safer, yet several large balloons of this class in Germany and France have met with disaster, involving loss of lives. The capacity of the dirigible for longer flights and its superior facilities for carrying apparatus and operators for wireless telegraphy are distinct advantages.

There has not yet been much opportunity to test the airship in actual warfare. The aeroplane has been used by the Italians in Tripoli for scouting and reconnoitering and is said to have justified expectations. On several occasions the Italian military aviators followed the movements of the enemy, in one instance as far as forty miles inland. At the time of the attack by the Turks a skillful aeroplane reconnaissance revealed the approach of a large Turkish force, believed to be at the time sixty miles away in the mountains.

Aeroplanes and airships, as they exist today, would doubtless render very valuable service in a time of war, both over land and water, in scouting, reconnoitering, carrying dispatches, and as some experts believe, in locating submarines and mines placed by the enemy in channels of exits from ports. A "coast aeroplane" could fly out 30 or 40 miles from land, and rising to a great height, descry any hostile ships on the distant horizon, observe their number, strength, formation and direction, and return within two hours with a report to obtain which would require several swift torpedo-boat destroyers and a much greater time. The question as to whether it would be practicable to bombard an enemy on land or sea with explosive bombs dropped or discharged from flying machines or airships, is one which is much discussed but hardly yet determined.

Aeroplanes have been constructed with floats in the place of runners and several attempts have been made, in some cases successfully, to light with them on and to rise from the water. Mr. Curtiss did this at San Francisco, in January, 1911. Attempts have also been made with the aeroplane to alight on and to take flight from the deck of a warship. Toward the end of 1910 Aviator Ely flew to land from the cruiser Birmingham, and in January, 1911, he flew from land and alighted on the cruiser Pennsylvania. But in these cases special arrangements were made which would be hardly practicable in a time of actual war.

In November, 1911, a test was made at Newport, R. I., by Lieut. Rodgers, of the navy, of a "hydro-areoplane" as an auxiliary to a battleship. The idea of the test was to alight alongside of the ship, hoist the machine aboard, put out to sea and launch the machine again with the use of a crane. Lieut. Rodgers came down smoothly alongside the Ohio, his machine was easily drawn aboard with a crane, and the Ohio steamed down to the open sea, where it was blowing half a gale. But, owing to the misjudgment of the ship's headway, one of the wings of the machine when it struck the water after being released from the crane, went under the water and was snapped off. Lieut. Rodgers was convinced that this method was too risky and that some other must be devised.

CHAPTER XXVIII. GLOSSARY OF AERONAUTICAL TERMS.

Aerodrome. — Literally a machine that runs in the air. Aerofoil. — The advancing transverse section of an aeroplane.

Aeroplane. — A flying machine of the glider pattern, used in contra-distinction to a dirigible balloon.

Aeronaut. — A person who travels in the air.

Aerostat. — A machine sustaining weight in the air. A balloon is an aerostat.

Aerostatic. — Pertaining to suspension in the air; the art of aerial navigation.

Ailerons. — Small stabilizing planes attached to the main planes to assist in preserving equilibrium.

Angle of Incidence. — Angle formed by making comparison with a perpendicular line or body.

Angle of Inclination. — Angle at which a flying machine rises. This angle, like that of incidence, is obtained by comparison with an upright, or perpendicular line.

Auxiliary Planes. — Minor plane surfaces, used in conjunction with the main planes for stabilizing purposes.

Biplane. — A flying-machine of the glider type with two surface planes.

Blade Twist. — The angle of twist or curvature on a propeller blade.

Cambered. — Curve or arch in plane, or wing from port to starboard.

Chassis. — The under framework of a flying machine; the framework of the lower plane.

Control. — System by which the rudders and stabilizing planes are manipulated.

Dihedral. — Having two sides and set at an angle, like dihedral planes, or dihedral propeller blades.

Dirigible.—Obedient to a rudder; something that may be steered or directed.

Helicopter.—Flying machine the lifting power of which is furnished by vertical propellers.

Lateral Curvature.—Parabolic form in a transverse direction.

Lateral Equilibrium or Stability.—Maintenance of the machine on an even keel transversely. If the lateral equilibrium is perfect the extreme ends of the machine will be on a dead level.

Longitudinal Equilibrium or Stability.—Maintenance of the machine on an even keel from front to rear.

Monoplane.—Flying machine with one supporting, or surface plane.

Multiplane.—Flying machine with more than three surface planes.

Ornithopter.—Flying machine with movable bird-like wings.

Parabolic Curves.—Having the form of a parabola—a conic section.

Pitch of Propeller Blade.—See "Twist."

Ribs.—The pieces over which the cloth covering is stretched.

Spread.—The distance from end to end of the main surface; the transverse dimension.

Stanchions.—Upright pieces connecting the upper and lower frames.

Struts.—The pieces which hold together longitudinally the main frame beams.

Superposed.—Placed one over another.

Surface Area.—The amount of cloth-covered supporting surface which furnishes the sustaining quality.

Sustentation.—Suspension in the air. Power of sustentation; the quality of sustaining a weight in the air.

Triplane.—Flying machine with three surface planes.

Thrust of Propeller.—Power with which the blades displace the air.

Width.—The distance from the front to the rear edge of a flying machine.

Wind Pressure.—The force exerted by the wind when a body is moving against it. There is always more or less wind pressure, even in a calm.

Wing Tips.—The extreme ends of the main surface planes. Sometimes these are movable parts of the main planes, and sometimes separate auxiliary planes.

Footnotes:

1 (return)
[Now dead.]
2 (return)
[Aeronautics.]
3 (return)
[See Chapter XXV.]
4 (return)
[The Wrights' new machine weighs only 900 pounds.]
5 (return)
[Aeronautics.]

www.ingramcontent.com/pod-product-compliance
Lightning Source LLC
Chambersburg PA
CBHW031631210526
45464CB00004B/1851